WILD DOGS OF THE WORLD

"The only recent composite work on the popular dog family. . . . This interesting study will provide the non-scientist with some insight that will help eliminate man's unjustified prejudice against predators."—*Library Journal*

"Well written. . . . Good reading. . . . An excellent reference source."—*Nashville Banner*

"A fancier's manual."—*Kirkus Reviews*

"Well researched. . . . Highly readable. . . . Rewarding."
—*Chattanooga Times*

"A great source of information, attractive in format and popularly written. . . . Worth the price as a settler of bets."
—Roger Caras

WILD DOGS OF THE WORLD

Lois E. Bueler

A SCARBOROUGH BOOK

SB STEIN AND DAY/*Publishers*/New York

FIRST SCARBOROUGH BOOKS EDITION 1980

Wild Dogs of the World was originally published in hardcover by Stein and Day/*Publishers.*

Library of Congress Catalog Card No. 72-96435

Designed by Bernard Schleifer
Printed in the United States of America
Stein and Day/*Publishers*/Scarborough House, Briarcliff Manor, N.Y. 10510
ISBN 0-8128-6075-6

Excerpts from Lois Crisler, *Arctic Wild,* used by permission of Harper & Row, Publishers; excerpt from Archer B. Gilfillian, *Sheep,* as quoted in J. Frank Dobie, *The Voice of the Coyote,* quoted by permission of Little, Brown and Co.; excerpt from Louis S. B. Leakey, *The Wild Realm: Animals of East Africa,* used by permission of The National Geographic Society.

FOR MY FAMILY

Contents

THE FAMILY 15

THE GENUS *CANIS* 19

Gray Wolf 22

Red Wolf 67

Coyote 72

Golden Jackal 83

Black-backed Jackal 89

Side-striped Jackal 97

Dingo 99

Domestic Dog 107

SEMYEN FOX 120

THE GENUS *VULPES* 124

Red Fox 127

Kit Fox 153

Corsac Fox 157

Bengal Fox 161

Hoary Fox 163
Tibetan Sand Fox 163
Rüppell's Sand Fox 165
African Sand Fox 168
Cape Fox 169

ARCTIC FOX 173

FENNEC 182

GRAY FOX 186

THE GENUS *DUSICYON* 191
CULPEO 193
Santa Elena Fox 195
PAMPAS FOX 195
Peruvian Fox 199
CHILLA 199
Chiloé Fox 201
FIELD FOX 202
Sechura Fox 203

FOREST FOX 204

SMALL-EARED DOG 208

MANED WOLF 212

RACCOON DOG 217

AFRICAN HUNTING DOG 224

DHOLE 237

BUSH DOG 243

BAT-EARED FOX 247

APPENDIX I—ANATOMICAL CHARTS OF THE DOMESTIC DOG 251

Bone structure of the domestic dog 251

Chief parts of the body 252

Skull seen from the side 253

Skull seen from above 254

The bite 255

APPENDIX II—SCIENTIFIC NAMES AND TAXONOMIC STATUS 256

Common and scientific names of all species of dogs 256

Taxonomic status and nomenclature of some species 257

APPENDIX III—RELATIONSHIPS AMONG MEMBERS OF THE DOG FAMILY 260

APPENDIX IV—GENETIC INFORMATION 262

Chromosome numbers of some species 262

Inter-specific crosses among dogs 262

BIBLIOGRAPHY 265

INDEX 271

List of Illustrations

1. *Moose standing its ground against attack by gray wolves* 35
2. *Gray wolf howling* 46
3. *Adult gray wolf regurgitating food for a pup* 55
4. *Red wolf, typical tawny color* 69
5. *Black phase of the red wolf* 69
6. *Coyote* 74
7. *Golden jackal with fellow-scavengers* 87
8. *Black-backed jackals feeding on zebra foal* 92
9. *Side-striped jackal* 98
10. *Dingo barking* 101
11. *Female dingo and pup* 105
12. *Semyen fox* 122
13. *Red fox and pups* 150
14. *Kit fox* 154
15. *Corsac fox with pups* 160
16. *Tibetan sand fox* 165
17. *Rüppell's sand fox* 167
18. *African sand foxes* 170
19. *Cape fox* 170
20. *Arctic fox in winter-white coat* 181
21. *Fennecs* 183
22. *Gray fox* 187
23. *Culpeo* 194
24. *Pampas fox* 196
25. *Chilla* 200
26. *Forest foxes* 205
27. *Small-eared dog* 209
28. *Maned wolf* 213
29. *Raccoon dog* 218
30. *African hunting dogs* 226
31. *African hunting dogs and pups* 232
32. *Dhole* 238
33. *Bush dog* 244
34. *Bush dog pups* 246
35. *Bat-eared fox* 248

Acknowledgments

THE FOLLOWING PEOPLE, among many, went out of their way to be helpful to me: Dr. John L. Paradiso read and criticized an early draft of the red wolf chapter and he, Dr. Philip W. Ogilvie, and Dr. Douglas H. Pimlott supplied material on gray and red wolves. Dr. R. Knox Jones answered my questions about the relationships of the dog family as a whole. Mr. R. Marlin Perkins supplied background for his photograph of the Tibetan sand fox, Mr. Kojo Tanaka filled in information on the raccoon dog in Japan, and Dr. David L. Harrison answered questions about Rüppell's sand fox. The Office of the Superintendent of Stock Routes, Department of Lands, Queensland responded at length to my queries about dingo control in Australia. Information on the Semyen fox was provided by letters from Dr. Jean Dorst and Messrs. Lawrence R. Guth, John H. Blower, and Leslie Brown. Dr. M. Kraus of the Nürnberg Zoo described in detail the zoo's experience with African sand foxes. And my father, Dr. Theodore H. Eaton, provided a great deal of material, much helpful criticism, and, above all, encouragement.

The Family

DOGS ARE THE ANIMALS most people know best. Probably the first animals to be domesticated, they have been our companions throughout recorded history. We know a great deal about domestic dogs: how they have been bred for special physical and psychological characteristics, how they are trained for specific duties, how best to make them useful and pleasant pets. But domestic dogs represent only a very small part of a large and widespread family which is much less well known to the general public. Everyone has heard of gray wolves and red foxes, of coyotes and dingos and jackals. But few people have ever heard of the Semyen fox of Ethiopia, or the raccoon dog of east Asia, or the maned wolf of Brazil.

Many wild members of the dog family are not only little known to the public, but have been little studied by scientists. Hard to observe in the wild, they do not yield the most interesting of their secrets in captivity. Wild dogs are shy, elusive, inconspicuous; some are nocturnal in their habits; many live in inaccessible places. Often they travel fast and for long distances, which makes them hard to track down.

This book is an effort to bring together information on all members of the large and important family of man's best friend. I have made no attempt to cover what is known about domestic dogs nor have I included all details available on the better known wild dogs. I have, however, tried to incorporate as much as possible about the more obscure members of the family—in some cases pitifully little. A great deal of wild-dog research remains to be done; until recently such work was merely descriptive of pelts and skeletons, with little field observation. In the past ten years the wolves of North America and the hunting dogs of Africa, among others, have begun to receive the attention they deserve. Until the other wild dogs are as well known, this book may serve as a collection of our fragmented knowledge.

What are dogs? What distinguishes them from other medium-sized carnivores? At first glance, their general appearance and posture. Members of the family usually have fairly long legs on which they stand upright, rather than crouching or slinking. All dogs, with the exception of a few domestic ones, are covered with hair or fur, sometimes thick and luxurious. Their color ranges from black to tawny or white, but only a few members of the family have any stripes or spots. Their ears stand erect, except for the floppy-eared breeds of domestic dogs, and may be either large or small, pointed or rounded. Their muzzles—again with the exception of some domestic breeds—are fairly long.

Dogs' feet are distinctive: with a single exception, they have five toes on the front feet and four on the hind feet, with long nails which cannot be retracted. The fifth toe on the forefoot is small and located above the pad on the inside of the leg; this is the dewclaw or pollex. Members of the dog family usually have long tails, always with some hair on them. Tail hair may be coarse or sparse, though foxes have dense and luxuriously furred brushes. Wild dogs tend to carry their heads parallel to or higher than their backs, but their tails are usually parallel to or lower than their backs. Some of these characteristics, notably those of the nails and the upstanding posture, distinguish dogs externally from members of the cat family.

Dogs have distinctive teeth. The incisors are small and unspecialized, but the canines, those four large ripping teeth at the front corners of the mouth, are sharply pointed and serve both to kill and tear apart prey. Behind each canine tooth is a row of small premolars. The last premolar on the upper jaw and the first molar on the lower jaw are called the carnassial teeth; they are especially large and also serve to rip and tear. The posterior molars, however, have grinding surfaces for chewing. Most dogs have, in their upper jaws, six incisors, two canines, eight premolars, and four molars. The lower jaw is the same except for two extra molars—making a total of 42 teeth. Only a few kinds of dog have a differing number of teeth.

Dogs are predators, though some also scavenge or eat vegetable food. Often they hunt in pairs or small family groups, though wolves and some other wild dogs form packs, while some species tend to be solitary. Dogs usually produce one litter of pups a year, numbering from two or three to a dozen or more. All dogs take care of their young for at least several weeks, some for several months. This long period of care provides for young animals that cannot hunt for themselves and also serves to cement social ties and behavior.

In many ways, dogs are the most generalized of all the carnivores, and it is this that accounts for their success as a family. Too much specialization is dangerous to an animal when the environment is in flux, especially as a result of the rapid and drastic changes made by man. If an animal absolutely requires a particular kind of climate, terrain, or food to survive, and that factor changes or disappears, the animal is doomed. But members of the dog family require little in their food or environment that is highly special. When their world changes they have often proved able to change with it.

Wild dogs lived originally on every continent except Antarctica and Australia. Elsewhere they were absent only on certain islands, including New Zealand, New Guinea, Melanesia, Polynesia, the Philippines, the Moluccas, Borneo, Formosa, and Madagascar. Though not indigenous to Australia the wild dog came there early: the dingo, which we now consider wild, was almost certainly brought to Australia by primitive man.

The domestic dog, of course, has been taken by man to all parts of the world but there are wild dogs living high on the plateaus of Tibet, in extreme desert conditions in the Sahara, in the jungles of the Amazon basin, and wherever there are forests and plains. One dog, the arctic fox, lives comfortably in the far north (he sometimes even makes long trips on ice floes); others, such as the red fox, live in suburbia and on the fringes of big cities. Although they are basically carnivores, some species of dogs will eat anything, from watermelons to bits of undigested food found in cattle and horse dung. Some dogs are excellent swimmers—the South American bush dog even dives under water; one, the American gray fox, climbs trees. Mountaineering history records at least one accomplished climber among domestic dogs, while the African hunting dog has been photographed on the summit crater of Mount Kilimanjaro at a height of over 19,000 feet. The animal world holds no more versatile creature than the dog—wild or domestic.

The domestic dog's relationship with man is a long and symbiotic one. People use dogs in an incredible variety of ways; the many specialized breeds would not exist without the intervention of man. But wild dogs are also important to man. The fur of some species, notably the arctic and red foxes, the raccoon dog and, to a lesser extent, the gray wolf, is valuable. Many species are inveterate rodent catchers and, where they are not persecuted, help to keep rodents in check. Also, as highly efficient predators, wild dogs are important in maintaining a balance among game herds in places where grazing animals still abound. Yet man has long been antagonistic towards many species of wild dog because he regards them as a menace to livestock and poultry and, when rabid, directly to man himself.

The Genus Canis

THE ANIMALS of the genus *Canis* are sometimes called the "typical" dogs. Actually, they are no more typical than are foxes or any other group within the family. But domestic dogs, along with wolves, jackals, and dingos, are the animals which man first recognized as dogs, and those on which the scientific world first based its description of the family as a whole. Originally a genus of the northern temperate zone, wild *Canis* today is also found throughout Africa and Australia. Domestic dogs, of course, are found nearly everywhere.

The wild dogs classified as *Canis* have a great deal in common, including a close genetic relationship. They are all medium- to large-sized animals, with relatively strong teeth and jaws, a round pupil to the eye, a gestation period of about 63 days and eight or ten mammae. All are capable of hunting fairly large game, though the gray wolf feeds almost exclusively on larger herbivores while smaller members of the genus depend more on rabbits, rodents, vegetable matter, and carrion. All are adapted to a variety of habitats, including open plains and tundra, forest, woodland, and semidesert. Though they can tolerate wide variations in temperature, none is found in the most severe desert conditions, probably because they are too large to survive on a diet of the very small mammals, reptiles, and insects on which desert predators typically depend. All are more or less social, the wolves forming packs while the other members of the family associate in pairs and small family groups. Though the members of this genus use a den to bear and raise pups, none is bound continually to a den the year around. The pups are well taken care of and usually associate with the parents during adolescence.

The gray wolf, most famous wild member of the genus, is the species with the widest original range. It was once found throughout the arctic and temperate zones of the northern hemisphere, though it is now gone from much of this original range and within a few years may occur in significant numbers only in Canada and Alaska. Ancestor of the domestic dog, the gray wolf is the largest of all wild dogs and one of the most highly social.

North America has two other wild representatives of the genus *Canis*. The coyote is a smaller, wolf-like animal, better adapted than its bigger cousin to a diet of rodents, rabbits, and vegetable food, though it sometimes preys upon larger game. Unlike the wolf, the coyote has learned to live with man. Although ruthless persecution has significantly reduced its numbers in some parts of the United States, in other areas it has spread to regions which it did not populate before the coming of white men.

The third North American member of the genus is the red wolf, the scientific classification of which is a subject of much discussion. Its original range was relatively small: the southeastern and southcentral United States from Florida west to Texas. Intermediate in size between the coyote and the gray wolf, the red wolf resembles the coyote in general appearance but, like the larger wolf, has shown itself unable to cope with a changing habitat and persecution by man and is virtually extinct. A curious fact is the recent extensive hybridization, in the wild, of the red wolf with the coyote, a situation which has helped cause the confusion about the wolf's taxonomic status.

The Old World, in addition to being a home of the gray wolf, includes the three members of *Canis* which we call jackals. The jackals are all similar in size, appearance, and habits. Considerably smaller than the wolves, they are not even quite as large as the American coyote. They can kill small antelope and deer and the new-born of the larger herbivores, but they also depend heavily on a diet of rodents, insects, and carrion. Jackals differ from wolves and coyotes in their voices, as well as in having somewhat smaller teeth and slightly different proportions of the skull. Less social than wolves, they live at most in pairs and small family groups. The black-backed and side-striped jackals are restricted to Africa, while the golden jackal is found from northeastern Africa throughout much of southern Asia.

We do not know very much about the origins of the dingo except that it is not indigenous to Australia and presumably was brought there by primitive man. It is very similar to some of the pariah dogs which roam south Asia; both dingos and pariahs may well represent an intermediate type between wolves and domesticated dogs. Dingos can easily be tamed when captured young—Australian aborigines use them today as hunting companions.

The domestic dog is, of course, the best known member of the genus. All domestic dogs are considered to belong to the same species and to be descended from a common ancestor: the gray wolf or possibly a now extinct close relative. The salient feature of the domestic dog, aside from its close association with man, is its great variation from breed to breed. Some members of the species have features not found in any wild species of the genus—hairlessness, for instance, or a snub nose, short or deformed legs, extremely long and floppy ears, very small size, distinctive voice, or specialized psychological and physical adaptations to such tasks as pointing birds, following a ground scent, or herding livestock. These features are the result of human intervention.

Physical anomalies such as the snub nose of the bulldog, the short legs of the dachshund, or the baldness of the Mexican hairless, all sometimes occur in wild species as mutations. For obvious reasons they are not advantageous to wild dogs and so disappear. But since man finds these mutations desirable, he breeds animals which pass on the traits to their offspring. In psychological adaptations such as an interest in herding sheep or in specialized hunting, he has strengthened a trait which wild members of the genus have to a less intensive or exclusive extent. Wild dogs, especially those of the genus *Canis,* must be able to do many things—they must be able to use both eyes and nose while hunting, they must be fleet of foot and capable of stealth, and they must be able to catch many different kinds of prey. But man has in some cases bred domestic dogs to hunt only by sight or only by scent, to hunt noisily or silently, to run very fast or barely to run at all, as it has suited his purposes. He has, in other words, produced a whole range of specialists from those superb generalists, the wild members of the genus.

THE GRAY WOLF (*Canis lupis*)

With its intelligence and intricate social structure, the gray wolf is one of the most fascinating of the world's great carnivores. Its wide distribution and long rivalry with man the hunter and stock-raiser have caused it to be an object of fear and fable perhaps unequalled by any other wild animal. Within its original range—much of the temperate northern half of the globe—it was once the most important predator, as the great cats are in the tropics. The wolf is losing its battle with man; its range is now severely restricted, its numbers diminishing rapidly. But it has not disappeared, and in the last few years man's fascination has begun to overcome his fear, so that the gray wolf is now the most studied and best known wild member of the dog family.

It has many regional names: timber wolf, European wolf, northern wolf, and lobo, but the single word "wolf" identifies it in most parts of its range. In the New World it was found from northern Greenland south to the high plateau of central Mexico, and coast to coast from Newfoundland to California. It was rather evenly distributed, with the exception of a few extremely arid regious in the southern and western United States, and the Baja California peninsula of Mexico. In the Old World, the animal's territory once stretched from Britain south to Spain and Portugal, across all of Europe, throughout Russia to northern Korea, Japan, and the Kuriles, and south through China to northern India. It occurred throughout the Near East as far west as Egypt. Among wild dogs only the red fox had as immense a range.

Except in very sparsely inhabited places the wolf is now, as mentioned, extinct or disappearing, and in many areas where it does remain, such as the Soviet Union, it is under heavy attack by man. An important factor in its decline is direct hunting, but a greater one is the decline of its own prey. Large hoofed animals can generally live only in vast and undisturbed areas, and wolves cannot live without these hoofed beasts—the ungulates. In a few cases, such as that of the white-tailed deer in the United States, man protects and encourages, largely for sporting purposes, an ungulate which can adapt to the mixed environment of forest and wooded farmland. But the wolf cannot safely follow the white-tailed deer into such a man-controlled environment.

The gray wolf is much larger than other wild dogs, but its size varies widely between individuals and in differing parts of its range. Females are usually smaller than males. The length of head and body varies from 41 to 48 inches. The tail is from a foot long to as much as 20 inches. Small wolves stand about 26 inches high at the shoulder, while the largest reach 38 inches. Small females and some animals of both sexes from southern populations weigh 60 pounds or less, while large northern males can weigh from 100 to 175. (The known record weight for a wolf is 197 pounds—it was a male killed in the foothills of Mt. McKinley in Alaska.) To give an idea of comparable size: a large, well-proportioned male German Shepherd stands 28 or 29 inches at the shoulder and weighs a little over 100 pounds.

Typical wolf color is difficult to describe. The basic types are black, white, and gray, and even within a fairly small population variation may be considerable. Arctic wolves, for instance, are often white or whitish, but all shades of gray or buff are found among them, and there are even blackish ones. Nor is the color uniform over the body. Gray wolves are gray because of the mixture of tones in each individual hair, which is usually dark near the roots, light in the center, and often black-tipped. Where guard hairs are thickest or longest, as on the shoulders and back, color is darkest. When a wolf molts and loses these guard hairs, the color fades and becomes more uniform over the body. In dark animals, the belly is usually lighter than the back. Some wolves have clear patterns—a black mantle over the back and neck of a whitish animal, a silver mane or ruff on a gray one, a dark band on the back of a light wolf, or a vertical light line back of the shoulders of a black one.

Color variation among individuals changes from one subspecies to another. Northern and tundra wolves may often be distinguished from each other on the basis of color alone, while on Isle Royale in Michigan and in Algonquin Park in Ontario, observers in low-flying planes have not been able to see a color difference between either packs or individuals. Within even one family it may vary widely. A litter taken from a den in Colorado contained two gray pups, one grayish-black, two nearly white, one grayish-buff, and the one brownish-black. So while for the sake of convenience we call this animal the gray wolf, we must remember that it is often not gray at all.

Size and color may be the most easily described characteristics of the wolf but they do not always distinguish an individual wolf from other similar dogs. Size is distinctive of the species as a whole, but some wolves are no bigger than large coyotes. To make identification more exact, researchers have in the past relied on a number of distinctive external characteristics. These include the proportions of the toes in relation to the whole paw, the width of the nose pad, which in a wolf usually exceeds one inch, and the carriage of the tail, which is usually stretched behind the animal as it runs.

None of these indicators is completely reliable. What is diagnostic, however, is the combination of the proportions of the skull: they are sufficiently distinctive to separate wolves from the animals that most resemble them—coyotes and wolf-like domestic dogs. Compared with coyotes, wolves and wolf-like domestic dogs have a relatively small brain case and large jaws. In distinguishing between the latter two, tooth size is the important feature; wolves have more massive teeth than domestic dogs of the same size.

None of this really tells someone who has never seen a wolf just what one looks like. At first glance it seems rather like a large, thick-coated German Shepherd. But a closer look reveals other characteristics: a wolf seems to have longer legs than a German Shepherd, a much narrower, deeper chest, a more slender and graceful body. Particularly in summer, when the hair is short, its ears are very prominent. The paws are long, with slender, spreading toes, and the tail, which is never carried above the back, often seems to float out and down.

Wolves have a freer range of movement in their limbs and tail than do German Shepherds, say. They will lay both tail and front and hind legs quite easily over each others' backs. Wolves fighting or play-fighting attempt to force down each others' forequarters with a fore-paw, or even to administer a "check" with a hind leg, sometimes pushing the opponent off balance. Although captive wolves do not easily learn to sit upon command, for they naturally sit far less frequently than dogs, "shaking hands" is second nature to them.

The wolf's usual gait is a trot so highly developed that it reaches probably the ultimate in efficiency among trotting dogs. The animal has narrow shoulders and well-turned-in elbows. The fore and hind feet on each side swing in the same plane, whereas domestic dogs are usually either wider or narrower in front than behind. The wolf's foreleg reaches far forward and the stride at its maximum is about equal to the height of the animal at the shoulder. This smooth, stretching, relatively effortless trot allows the animal to cover as much as 20 or 30 miles in a single outing.

Wolves also lope—not fast, but tirelessly. They have been clocked at a speed of about 28 miles per hour over a distance of 200 yards, and can probably average from 22 to 24 miles per hour over the first mile or two of a run, not very fast as running animals go. After this they slow down as much as fifty percent, but can continue to lope for hours.

The tracks of a very large wolf are distinctive, but those of average-sized wolves are hard to tell from the tracks of big dogs. Expert trackers know that the front two toes of the wolf are often closer together than those of the dog. The prints of the nails are more prominent, too, a feature which can usually be seen only in dry soft earth. The wolf's track is somewhat longer and narrower than a dog's—a rough trapezoid rather than the dog's more nearly circular print. The fore feet of the wolf are distinctly larger than the hind feet. The size of the tracks will depend on the gait of the animal, for when the wolf is trotting or loping the feet spread considerably. It is often said that the tracks of a wolf fall in a single line, and that groups of wolves follow each other in single file. This may be true in soft deep snow, when traveling like this would reduce the strain on each individual limb, and on the pack members which follow the leader. In other conditions, however, the hind feet may or may not fall in the prints of the fore feet, and animals in a group may spread out as they travel.

Wolves have been divided into a great number of subspecies, 23 or 24 in North America alone. Four subspecies have been described for Europe and the Soviet Union, and two more for the Indian subcontinent. The wolves of southern Japan and those of the Near East have also been accorded subspecific status. Subspecific distinctions have been based chiefly on gross size, general color and color patterns, and skull and tooth measurements. But many of the older subspecific divisions among the dog family are no longer considered valid. Wolves found in a given region may indeed differ from those of another region in details of size, color, or proportions, yet intergrade in such a way that the two cannot be separated meaningfully except at the extremes of each range.

Though the gray wolf once lived throughout most of temperate North America, the arrival of Europeans has steadily reduced its range. Today, outside of Alaska, wolves are virtually extinct in the U. S. Most of the remnant populations which still existed in parts of the west even a few decades ago have now gone. There is a significant wolf population in northeastern Minnesota—estimated at 250 in 1956—a few in small sections of Michigan's upper peninsula and possibly in northern Wisconsin, and a pack of about 25 in Isle Royale National Park in Lake Superior. This small Isle Royale group is far more important to science than its numbers indicate because, since it lives in the naturally circumscribed environment of an island, it has been the object of years of intensive study. Another minor but encouraging sign is the apparent survival of a few wolves in Yellowstone Park. Systematic monitoring and recording of wolf sightings there indicate that perhaps a dozen wolves live in the Yellowstone area. The policy of Park authorities is to encourage the presence of wolves, although they are subject to poisoning and hunting if they leave the Park itself. Yellowstone could some day support a viable wolf population but the day has not yet come.

Alaska and Canada are the strongholds of the wolf in North America, though in Mexico it still hangs on in the Sierra Madre

Occidental and in the mountains of eastern Chihuahua and western Coahuila. Until recently Alaska offered a $50 bounty on wolves, and the predator was widely hunted by airplane. Fortunately the bounty is no longer in effect, and hunting animals from aircraft is now illegal under federal law. Most of Canada has discontinued the bounty system but controls are exercised by various provincial government agencies. Methods include wide use of poison stations and even, at times, dropping poisoned bait from the air. Now, however, wolves are no longer being poisoned in many wilderness areas and pressure against them has slackened.

The virtual extinction of wolves in the United States, outside Alaska, is due largely to the slaughter of great numbers in the 1800's, mostly by strychnine poisoning, and to continued hunting ever since. But the disappearance of prey has been more important in their decline elsewhere. For instance, several severe ice storms caused widespread starvation of caribou around Thule in Greenland, and by 1900 all had disappeared. The wolves which had preyed on them became scarcer and finally disappeared completely; the last wolf was reported in eastern Greenland in 1939. Curiously, there seems no reason for the disappearance of the wolf in Newfoundland. These animals had probably colonized the great island by drifting south from the Arctic on ice floes. They were apparently not extensively hunted by man, no great change occurred in the habitat, and caribou remained in good numbers, yet the wolves had disappeared by early in the 20th century.

In Europe, wolves disappeared as they have in the populated areas of the U. S.—killed off by hunting and by the spread of intensive agriculture. Ireland and England were once the animal's strongholds, but its decline began there hundreds of years ago. Both in Roman times and after the Norman conquest, levies of wolf pelts were exacted as rent or tribute from feudal landholders. A number of methods of hunting were used, from community wolf drives to the use of pits, snares, and even the breeding of the huge Irish wolfhound. Under such pressure, wolves became extinct in England by about 1500, in Scotland and Ireland by the end of the 18th century. They continued to be common in the mountains of Europe for another 100 years, but today the wolf in western Europe is on the verge of extinction. There are probably none left in France or Switzerland. In Italy a few still live in the Abruzzi Park in the Apennines and they are fairly common in parts of the Iberian peninsula.

The wolves of Scandinavia are concentrated largely in Finland, though there are a few in northern Sweden along the mountainous border with Norway, a country which also has a remnant population. The wolves of northern Scandinavia originally preyed on wild reindeer. Around the 9th century A.D., or even before, the Lapps began to domesticate reindeer, wolves turned to them as their quarry, and the Lapps began to wipe out the wolves. With the declining number of wolves in Lapland, depredations on the reindeer herds are less and less significant, but wolves continue to be hated because they scatter the herds. The Lapps of Sweden would like to see the wolves in that country exterminated altogether, while a few Swedish citizens, and international conservation groups, want them protected.

In eastern Europe, following a rapid decline in the first part of this century, there was an upsurge in wolf numbers during and after World War II. The war itself and the socialist governments which followed reduced the number of hunters who shot wolves for sport. After the war, game management and agriculture were put in the hands of the state and some farmland in areas strategic to wildlife returned to forest. For a while at least such changes favored an increase in game animals and attendant predators. But the governments of eastern European countries have responded with vigorous wolf control programs; in Poland, for instance, some 1,000 wolves were killed in 1960.

Besides Canada and Alaska, the Soviet Union is now the last stronghold of the wolf. There are few in European Russia, but thousands still exist in Siberia. Under present conditions they cannot remain for long, since Soviet policy vigorously supports a program of complete destruction of the wolf, on the grounds that it damages livestock and game animals. Wolves are exterminated however possible, mostly from airplanes with automatic weapons and by use of poison; the effectiveness of the onslaught can be measured by the drop in the annual kill—from 42,600 in 1946 to 8,800 in 1963. In the rest of its original range the wolf's existence is marginal. It is now gone from Japan and is scarce in

India. Though still widespread in the Arabian peninsula and much of the Near East it does not occur there in any great numbers.

Wolves have no significant natural enemies other than man. They are susceptible to accidents, especially the dangers involved in hunting large and sometimes formidable prey. They can be victims of severe weather as well. They shun extremely dry conditions, and sudden harsh blizzards have been known to kill them. Disease may be more important as a natural check than either weather or accidents. Wolves are affected by all the diseases and parasites common to domestic dogs, including rabies, mange, distemper, encephalitis, and worms. A reduction in wolf numbers that occurred in Alaska in the 1920's and 1930's may have been brought about by distemper introduced with the large numbers of sled dogs brought into the area then.

Rabies has occasionally been prevalent among wolves, especially in Eurasia. American Indians recognized the disease and feared it, for a rabid wolf, like any other animal afflicted with the disease, loses its fear of man and may enter encampments and attack dogs, livestock, and human beings.

Records of the 1870's from Fort Larned, an army post on the Arkansas River, detail the havoc of such a visit. "On the 5th of August, at 10 p.m., a rabid wolf, of the large grey species, came into the post and charged round most furiously. He entered the hospital and attacked Corporal ——, who was lying sick in bed, biting him severely in the left hand and right arm. The left little finger was nearly taken off. The wolf next dashed into a party of ladies and gentlemen sitting on Colonel ——'s porch and bit Lieut. —— severely in both legs. Leaving there he soon afterward attacked and bit Private —— in two places. This all occurred in an incredibly short space of time; and, although the above-mentioned were the only parties bitten, the animal left the marks of his presence in every quarter of the garrison. He moved with great rapidity, snapping at everything within his reach, tearing tents, window-curtains, bed-clothing, etc. in every direction. The sentinel at the guardhouse fired over the animal's back, while he ran between the man's legs. Finally he charged upon a sentinel at the haystack, and was killed by a well directed and most fortunate shot." (The corporal bitten at the hospital later died of hydrophobia, though the other two men who were bitten recovered from their wounds and never showed signs of the dread disease.) No wonder the cry of "mad wolf" has struck terror into the hearts of people throughout its range.

In addition to the typical canine diseases, mature and old wolves can be afflicted by cancer, and by arthritis, which seems to attack the legs and vertebrae. Such a disease must be especially crippling to an animal which hunts large prey over wide areas. Probably wolf pups die more often from heavy infestations of parasites than from any other ailment except distemper.

We now know a great deal about the behavior of wolves, as well as about their physical characteristics. They are found only where forage conditions will support their prey, which means the arctic tundra for caribou; the prairies and steppes for bison, pronghorn, Asian antelope, and some deer at certain times of the year; open mountain slopes for mountain sheep; secondary forest along rivers and lakes for moose and European elk; and mixed deciduous forest and meadows for most kinds of deer. The wolf is not found in any numbers in coniferous forests which contain little or no grazing for hoofed animals. Wolves generally avoid areas where there is very deep snow in winter; they themselves cannot move easily in deep soft snow and prey animals also move out of such areas to winter in more favorable terrain. The original distribution of the wolf was undoubtedly affected by the forest-cutting and livestock-raising activities of early man. As the forests of Europe and Asia were cut, some areas became more acceptable to grazing game than they had been before, and the wolf followed its prey. The presence of domestic cattle, sheep, and reindeer in large and unprotected herds introduced a source of easily obtainable food.

The nature of its prey has, in turn, determined the wolf's social structure. It is a highly organized social animal in a family which typically is composed of members with rather solitary habits. The structure of the wolf pack is ideally suited to the hunting of large prey. A single wolf, while it may kill a wounded or sick moose, normally has little hope of bringing down a healthy adult. A wolf pack, although it may have to test many animals before finding one which is vulnerable, is in a much better position to press its attack to the kill. The same is true even of smaller and less formidable prey like deer; several wolves can corner a deer, even in terrain covered with obstacles, where a single predator would have far less chance of success. Group hunting is especially important in raising the young, for a litter of half-grown pups requires a great deal of food, more than two parents can hope always to provide. The pack, too, can provide a sort of social security for old and injured members: sometimes these are allowed to remain with the pack and are in effect fed by it, but even if they are rejected and relegated to the position of camp-followers, they may still feed on prey abandoned by the main group. It is not surprising, therefore, that wolves have a host of mechanisms for maintaining social cohesion; mechanisms of voice and scent contact, patterns of dominance and submission, intricate patterns of juvenile behavior towards adults, and social as well as sexual bonds between mated pairs—all these develop and preserve wolf society.

There are still popular misconceptions of the wolf as a predator. Many people think that wolves and other large predators have only to go out on a fine morning, locate an innocent doe or helpless fawn, pounce upon it, and tear it to pieces for an instant meal. A corollary of this idea is that the predator commonly and willfully slaughters not just one, but numbers of helpless prey at a given time. People also assume that the sight or sound of a wolf invariably induces panic in all other animals. It may come as a surprise, therefore, to learn that under natural conditions wolves usually have to work very hard for a living. They may search out and test numerous animals for each individual they succeed in killing and they commonly hunt unsuccessfully for several days between kills. Moreover, each of their major prey species has its own very effective defense and prey animals, though wary of wolves, often casually accept them as part of their surroundings. All this is not so amazing, considering that were it otherwise wolves would quickly decimate their prey and leave themselves nothing to eat. As it is, the life of the predator may be free, but it is not soft.

Another common assumption is that wolves hunt all their prey the same way. In practice a successful hunt depends on successful tactics, and the tactics vary with the animal hunted.

There are constant factors on the wolf's side, of course; it has enormous endurance, great intelligence, sharp senses, powerful jaws, and large teeth. But these assets are applied in different ways against prey with different means of escape. The caribou runs away, and usually successfully. The moose may run, may stand at bay, or may even counterattack. It, too, is usually successful. The mountain sheep depends on favorable terrain for its defense against the wolf; if it can get above the wolf, especially on rough ground, it is virtually unattackable. The musk-ox has evolved a famous method of group defense, and a herd which stands in a circle with horns outward—and younger members inside—is invulnerable. Even the helpless fawn or calf has a defense in numbers, for fawns of a given species are dropped at the same time, and the relative bounty which the wolf may then enjoy, if it can locate the young and break through the defenses of the mother, is offset by the fact that this embarrassment of riches occurs only once a year and lasts for a short time. So even for the powerful wolf, survival means an unending battle of wits and strength against the evenhanded opposition of nature.

Like other dogs, a wolf will eat nearly anything. As the animal travels, it occasionally pauses to snap up rodents, hares, rabbits, marmots, and very small fry like lizards, snakes, tortoises, and insects. At a spot where fish come into shallow water, the wolf will wade in to scoop up or snap out a fish and then return to the bank to eat it, often beginning with the head. It takes advantage, too, of vulnerable birds like molting geese. Beavers are an important secondary source of food, chiefly during the spring and fall, when they travel as much as several hundred feet from water to cut trees, and can be surprised. Wolves sometimes surprise or even chase other wild dogs such as foxes and raccoon dogs, and have also been known to carry off domestic dogs. Where men are or have been, wolves raid garbage dumps and scavenge through debris. Though they usually kill their own prey, they have no objection to carrion of the right species. Especially in the southern parts of their range, wolves feed seasonally on wild and domestic fruit. But though they will eat nearly as many different kinds of food as coyotes or jackals do, the important difference is that wolves apparently cannot raise their litters on the smaller items alone. Even in parts of the tundra where caribou are highly migratory, and seem to desert parts of their range for a good portion of the year, enough stragglers remain for wolf populations to subsist.

Since adult wolves typically exist in a state of alternate feast and famine, they gorge when food is available and then endure long intervals between meals. An exhaustive study of the feeding habits of the wolves on Isle Royale determined that though each animal averaged perhaps 10 lbs. of food per day, it by no means consumed this amount regularly every day. Kills were made on the average of once every three days, but occasionally kills were made on successive days and at other times the pack went for five or six days without making a kill at all. To survive such an irregular regimen, wolves must be capable of eating a great deal at one time. On Isle Royale the largest pack, with 15 members, several times devoured a moose calf within 24 hours, at the rate of about 20 lbs. of food per animal. Elsewhere wolf stomachs weighing 18 and 19 lbs. have been recorded. Of course, the amount of meat which a wolf consumes from a fresh kill at one feeding is determined by the size of the pack and the size of the animal killed. A pack of several adults will quite often completely consume the kill at once, leaving only a few pieces of hide, the jaw and parts of the skull, and gnawed leg bones as the meager remains.

When wolves kill prey large enough to provide more than one meal, they normally eat to repletion, retire to a resting spot from several hundred yards to as much as several miles away, and then return after some hours to feed again. The idea that wolves do not return to a kill is probably based on the learned behavior of wolves that are much hunted. Where carcasses of wolf kills are poisoned by man, the wolf that survives is the one that moves on to kill again rather than returning to a previous kill. Normally wolves stay near their kill until it has been completely devoured; they prefer the dinner they are sure of to a meal which has not yet been captured. Carcasses are devoured especially rapidly when young pups must be fed, for they need a great deal of food and must eat more regularly than adults.

Wolves sometimes cache their food to protect it from scavengers or other wolves, or to feed it to their pups. After feeding, a wolf will carry a portion of its kill—the leg, shoulder, or head,

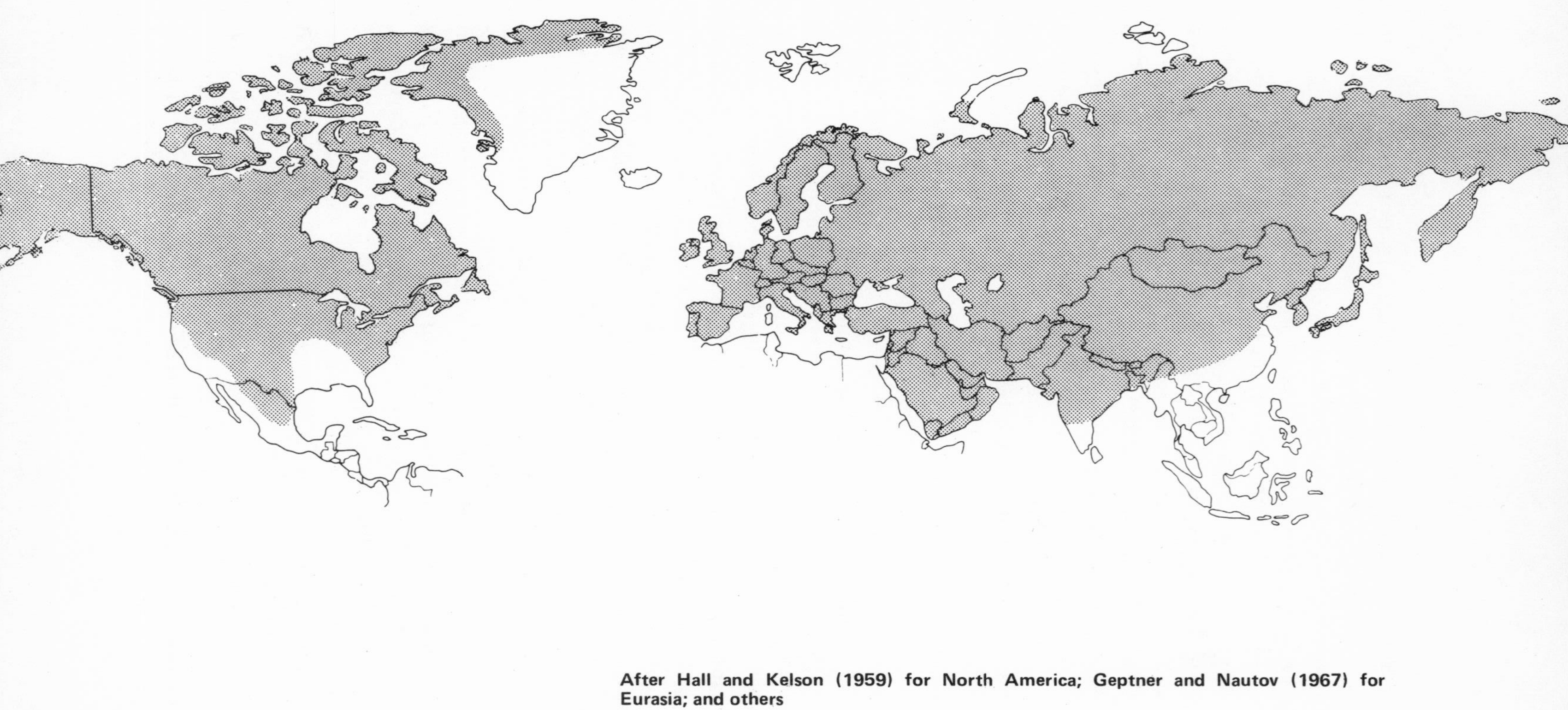

After Hall and Kelson (1959) for North America; Geptner and Nautov (1967) for Eurasia; and others

A moose stands its ground against attack by gray wolves. These wolves gave up and left after five minutes. *Photo: L. David Mech*

perhaps—to a spot where the ground is soft. It digs a shallow hole, deposits the meat, and pushes dirt over it with its nose. The cache is often made near the den and is visited not only by the adult which has stayed with the pups, but between kills by other members of the pack. Yet wolves often do not return to a cache, either from forgetfulness or because they have made a new kill. The beneficiaries then are bears, foxes, eagles, and other flesh-eaters that find the cache by scent or have seen the wolf going about its storing.

The most spectacular part of wolf life is hunting, and the most formidable prey is the moose. A healthy adult moose which stands at bay can be a match for an entire wolf pack and, unless its self-defense is hampered by deep or encrusted snow, it is practically invincible. Most attacks on moose, therefore, are processes of testing; only by pressing an attack can the wolves discover animals rendered vulnerable by old age, malnutrition, parasites, or disease. A moose at bay rears, wheels quickly, lashes out accurately with both fore and hind feet, will ferociously pound any predator it reaches, and will actually pursue wolves if sufficiently enraged. If possible it backs against a tree or brush to protect itself more easily. When a moose cow and calf stand together, the cow protects the calf's flanks and lashes out behind, while the calf charges forward. Wolves will rarely harass a standing moose for more than five minutes; it becomes simply a waste of their time and energy.

Moose do not always stand at bay, however. Sometimes they run immediately; at other times they turn and flee after a brief stand. If the moose bolts, the wolves follow it with great bounds, which they can keep up for as much as twenty minutes if the snow is not too deep to hamper them. Wolves and moose are about equally matched in speed, though uncrusted snow over two feet deep favors the long-legged ungulate, whereas in more shallow snow the wolves may quickly catch up. While wolves often abandon the pursuit of a running moose—particularly if it maintains a good lead for even a few seconds—the moose killed by wolves is usually the one that runs. A running animal cannot use its hoofs to full advantage. More important, weak animals are those most inclined to flee, and most likely to show their weakness during the chase, thus encouraging their pursuers to press the attack.

If the wolves can catch up with and attack the moose, the killing itself is fairly standardized. Several wolves bite and harry the flank of the animal, while one searches for a nose hold. The nose hold is not mortal, but it distracts the prey and keeps it from running. Once a wolf secures a nose hold, the rest of the pack rips into the rump, not because it contains any especially vital parts but because it is the safest area to attack. When the moose falls the wolves swarm over it and it is quickly killed. Small packs, or wolves attacking an especially large animal, may stop the attack and wait for the moose to stiffen and weaken. This is less a matter of conscious tactic than a result of the weariness of the wolves themselves, and their inability to bring down their prey. Occasionally wolves will abandon a wounded moose, which usually dies within a week and then may be found by the pack and eaten. Or a lone wolf, dependent on the leavings of the main pack, may take up its stand near the wounded animal and wait for it to die or weaken enough to be killed by a single predator.

Obviously wolves do not kill all the moose they encounter, or even most of them. The careful study of wolf predation on Isle Royale provides precise figures on the efficiency of wolves as hunters, and is well worth quoting. L. David Mech writes that "a total of 160 moose were estimated to be within the range of the hunting wolves while under observation, but only 131 were detected. Of these, 77 were tested by the wolves; i.e., the wolves chased them or held them at bay, so those which escaped did so because of their superior condition or ability. Those which were detected but not tested also may have escaped on this basis, but circumstances probably were more important. Therefore, predation efficiency is considered here as the percentage of animals tested that are killed. Since six moose were dispatched out of 77 tested, the predation efficiency is seven and eight-tenths percent." This surprisingly low figure of killing efficiency corresponds generally to that known for wolves in other situations, and for other types of predators.

Since so few animals are killed out of the total population of moose, it is obvious that the wolves must be selecting for inferiority of some kind. On Isle Royale, calves under a year old comprised about 15 per cent of the moose population but represented 36 per cent of the kills. They are vulnerable for the obvious reasons of inexperience, weakness, and dependence upon the mother, but also because of susceptibility of parasitism and disease. Almost all the rest of the moose killed were over six years old, the majority eight years or older. The most interesting fact is that moose between the ages of one and six years simply were not killed by wolves. These are the young adult moose, the most alert, healthiest, and most agile animals.

The wolves of Isle Royale feed almost exclusively on moose during the winter, and on moose supplemented by beaver in the summer, because they have no choice. The island does not contain any other animals large enough to support them. But in other parts of its range the wolf by preference hunts game other than moose, even when moose are also present. The major prey on the tundra, for instance, is caribou. Like moose, caribou have their own system of defense against wolves, in their case, sheer speed. Observers agree that a herd of caribou does not seem to be bothered by the proximity of wolves. It flees only if chased, so that wolves seldom stalk caribou and are able at will to approach only within a few hundred yards of them.

The caribou's defense of speed against wolves is so efficient that the predators must test many animals for each one they catch. They seem first to look over caribou herds for calves. If there are none, wolves will often not even bother to chase the herd. If they do begin to run the caribou, the chase may last for a long time. The wolves are watching for a weak calf that shows signs of faltering, or for the handicapped adult that starts to slow down. A large herd being chased by wolves often splits up into several groups. At this point the wolves usually pause, select a splinter group to pursue, and set off after it. Since healthy caribou, including young calves and cows about to calve, can outrun wolves with ease, the only animals that slow down enough to be caught are those suffering from broken legs or hoof disease, and those afflicted by such common caribou ailments as tapeworm cysts in the lungs, or swarms of botflies that clog the nostrils. It is interesting that being in a herd is often a disadvantage for a calf. It runs with the herd but is not always sure why it is running, and may not realize until too late that the animal pulling up beside it is a wolf and not another caribou. When it is by itself, however, the healthy calf senses enough danger simply to step on the accelerator and leave the wolves behind.

Mountain sheep, whether the bighorns of the Rockies, the Dall sheep of Alaska and northern Canada, or those of the Old World, have yet another means of coping with the threat of wolves. The rams, which have massive horns, may turn to face attacking wolves and batter them. But the favorite means of escape is flight, not out onto the plains, as with caribou, but up onto the rocks. Sheep have learned that if they can keep the predator in sight and can stay above it on rocky ground they are safe. When wolves chase them from below, the sheep, even week-old lambs, escape upward with little difficulty, while the wolves labor mightily up the same incline. Most wolves will not continue such a pursuit for long. When wolves do kill mountain sheep, the kill follows a drive down hill, or it occurs where the sheep can be surprised. Sometimes a particularly bad winter covers the high mountain slopes with wet crusted snow which the sheep cannot travel in or paw through. They are driven, starved and weak, down to the flatlands, where wolves can pounce at will.

The tactics wolves use against wild sheep seem to be much the same wherever the two animals range together. In the Himalayas and northern India, wolves seem to have little success against the goat-like blue sheep or bharal (*Ovis bharal*), which rarely leaves the rough terrain that is its favorite habitat. But they can kill the argali (*O. ammon*), especially when heavy snows hamper its movements.

Wolves hunt white-tailed deer about as successfully as they do other prey. Deer are smaller and thus less formidable than moose, mountain sheep, or caribou. But they typically inhabit woodland full of natural obstacles like brush, downed tree-trunks and rock outcroppings. With their high, bounding gait, they have a running advantage in such terrain. A wolf which starts a deer and gives chase must usually catch up with it quickly before the deer can use the terrain to its advantage. When heavy snows force deer together in yards for long periods of time, however, the steadily diminishing food supplies and the confinement of the yard put the deer very much at the mercy of wolves. Under such conditions a single wolf pack, traveling from yard to yard, can kill dozens of animals in a few weeks.

A pack can hunt deer more efficiently than a single wolf, but the advantage is probably the result of sheer numbers rather than strategy. For instance, wolves pursuing a deer often run strung out in a long irregular line. Deer which circle, as they commonly do, may cross their own back trail and be pulled down by the pack stragglers. Aerial observers in Minnesota have watched wolf packs which were crossing a frozen lake or river split up when they reached an island or spit of land, one group hunting across the land while the remainder of the pack circles on the ice. If a deer is sprung by the animals on land, those on the ice may be in an ideal position to pull it down as it flees. There is no way of proving that when a pack splits up and subsequently effects a kill it is actually practicing strategy. It may simply be that some animals prefer to travel across land and others are inclined to keep to the ice.

The great herds of plains bison and the enormous prairie wolves which followed them both disappeared, virtually together, before the onslaught of the white man. Early writers who had had the fortune to see bison in their heydey often described the wolves which loitered on the fringes of the bison masses, waiting for the chance to attack an unwary or debilitated animal. Other than man (both primitive and civilized), the bison's chief predator was the wolf, for of the large American carnivores it alone had the size, strength, and social organization to hunt these large hoofed animals systematically. Wolves haunted the calving grounds and did their most successful hunting when the calves were too weak to stay with the herd. Probably the chief safety of the calves lay in the sheer numbers born at the same time.

Old accounts indicate that wolves were highly respectful of the power of a group of adult bison. They attempted to locate weak or disabled animals, or if they themselves were in sufficient number, would try to cut a single bison from the main herd, harrying it away from the protection of the group until a mass attack could succeed. Writers often mentioned wolves attacking the animal's flank and nose, much as they do with moose. Since old or feeble bison soon fell to the rear and sides of a moving herd they were highly vulnerable to wolf predation. If a single animal, such as an old bull, was too strong to be brought down at once, the pack might harry its flanks to cause torn muscles and blood loss. Recent observations in Wood Buffalo Park in northern Canada, where bison form a staple of wolf diet both summer and winter, bear out the fact that wolves kill very old animals, calves, and injured or diseased adults. Young healthy adults are virtually invulnerable.

The effects of wolf predation on prey populations is only now beginning to be understood. In the past the average layman, as well as many biologists, thought that all predation was bad for the prey, and that the best way to protect and ensure the successful survival of herds was to wipe out the predators. That this idea dies hard, even now, is apparent from the attitude of Soviet authorities, who have embarked on a wolf extermination program not only to protect livestock but also, supposedly, to maintain wild game. On the other hand, since animals like the white-tailed deer in North America often seem to multiply so rapidly that they eat themselves out of forage despite the presence of wolves, some observers have come to the conclusion that wolves never did exercise a controlling effect on their prey.

Both these interpretations are called into question by current work on wolves. The fact that wolves ordinarily prey on the oldest and the youngest animals in a population means that their activities are in part compensatory. That is, in killing older animals they are killing those that would die anyway of other causes, or at least those which do not reproduce at maximum efficiency. Certainly some of the young animals are also destined to die anyway—of disease, starvation, or accident. On the other hand, the question of whether wolves can control prey populations is complicated by man's interference in the ecological picture. In North America, for instance, human activity has changed much deer and moose range in favor of these browsers, which thrive in areas cut over, burned, or once marginally farmed. That wolves may not be able to keep moose or deer numbers under control in such areas does not prove that under natural conditions, which exist only rarely in today's world, they might not have done so.

Exactly what constitutes a stable predator-prey relationship depends on conditions within each particular area and varies from prey species to prey species. A state of equilibrium is probably reached between wolves and moose on Isle Royale at a ratio of about one wolf to 30 moose. A ratio of one wolf to 100 white-tailed deer may be close to an equilibrium for that prey species in Ontario. Beyond those ratios, wolves are probably not capable, by themselves, of controlling prey populations. There is no reason to think that normal numbers of wolves harm stable populations of prey animals living under natural conditions; the effect of wolf predation under such circumstances is to maintain the prey in relatively healthy and well-balanced condition. But when man tampers with the habitat, or with either factor in the prey-predator equation, the delicate arrangement arrived at over millenia is thrown out of balance.

How many wolves may live in a given range? Most estimates have been sheer guesswork, like the estimate of a wolf population in primitive North America of two million animals, or one animal per 3½ square miles. The one uniform feature of such guesses has been considerable exaggeration. In fact, the highest density known to occur anywhere is on Isle Royale, where in 1965 there were 28 animals, a density of one wolf per 7½ square miles. In nearby northeastern Minnesota, 1955 estimates of the population were one wolf per 17 square miles. Studies in Algonquin Park resulted in an estimate of one wolf per ten square miles, but the wolf population in the rest of Ontario is probably closer to one wolf for every 100 to 200 square miles.

Many factors have led to the exaggerated estimates of the past. One is the mobility of the wolf; the same animal or group may be seen within a short period of time in places quite far from each other, and thus be counted twice. If wolves are seen frequently in some areas of a large range, it does not necessarily follow that more wolves are evenly dispersed in like numbers everywhere else, since the observer may be in a part of the range where wolves pass frequently. The only sure way to conduct a count is to be able to recognize individual animals, to use some system of live-trapping and tagging, or to perform an aerial survey. A safe rule of thumb is that in most places, except those such as Isle Royale where evidence is unusually clear and well documented, there are fewer wolves than is commonly thought.

Knowing how wolves hunt helps us explain many related aspects of their behavior, such as territorial and travel patterns. Since large game animals eat a lot of food they travel widely in search of forage, and often migrate from one favorable range to another. Wolves must move in search of their prey or starve. But wolf movements are not haphazard; individuals and packs inhabit territories and follow regular routes and trails. These routes almost always follow terrain where footing is good and the traveling easiest. Wolves may prefer old roads, game or livestock trails, narrows between bodies of water, and natural fords across streams; in short, roughly the same places where other large animals, or man on foot or horseback, would travel. In winter, wolves travel along frozen rivers, lakes, and streams and on open ridges where drifts are hard-packed. In places where the terrain restricts movements, the routes may be paths only a few feet wide, and may widen to a mile or more where the traveling wolves can range from side to side of a valley or open space. Wolf travel routes are often more or less oblong or circular, since following natural features, say up one river valley, across a ridge, then down another valley, may yield such a pattern. But there is little evidence that, as was once thought, wolves patrol a circular runway in a regular pattern; much less is there any truth to the notion that they move always in a counter-clockwise direction.

Wolves seem to spend about one-third of their time traveling, usually at a trot averaging about five m.p.h. Travel patterns are as uneven as feeding patterns. On Isle Royale a wolf pack's movements were followed consecutively for 31 days. During this time the pack traveled approximately 277 miles. On 22 of those days the wolves fed on kills and moved little. In nine days of actual traveling, the animals averaged 31 miles per day, and once traveled 45 miles in a single 24-hour period. Information about wolf travel patterns in other parts of the world—the U.S.S.R., Finland, Alaska—is similar, though less detailed.

The size of the pack territory or hunting range varies with topography, prey density, and the size of the wolf population. The 50-mile range of a pack studied in Alaska may be typical of northern wolves, but in Algonquin Park, packs hunt a territory no more than ten miles across. In summer the pack usually stays within the center of its range, expanding its movement in the fall, and in winter hunting to the limits of the pack territory.

There is still some mystery about how wolf packs delineate their territories. It has long been thought that scent posts are a primary means but there is little evidence that wolves systematically maintain scent markings to warn off their neighbors. Wolves sometimes concentrate their droppings in certain places but this may be simply because an animal passing another's feces often feels obliged to make his own contribution. Howling seems to be a means of territorial delineation in that it allows each pack or animal to know where its neighbors are. Wolves probably do not howl for the express purpose of marking boundaries, however. It is interesting that neighboring packs will often share a *de facto* no-man's land, a strip that animals of either pack may occasionally enter but which neither group includes in its regular travels.

A pack, as such, may consist of a pair of wolves with its offspring, or of two or more such family groups. Or it may involve a more irregular grouping of adults. Packs vary in size from area to area and according to the season; they are larger in winter and often split up into smaller groups in summer. In the Canadian Rockies, packs of five or six animals seem to be typical but there too, as elsewhere, groups of 15 or 20 may occur. Packs with as few as two or three animals are also seen frequently. Packs tend to be larger where the overall wolf population is high. On Isle Royale, an area of high population density, the largest pack consisted of about 15 animals and retained this size, at least in winter, for many years in succession.

Old accounts of wolves both in America and abroad include tales of packs that numbered hundreds. From all that is scientifically known about wolves, stories of such immense packs can be totally discounted. In the old west dozens of wolves might sometimes congregate where buffalo hunters were slaughtering game, but such congregations were not single packs. Communal howling may give an untrained ear the impression of scores of animals participating, whereas in truth there may be only a few. It is a matter of logic as well as observation that packs of great size are simply not efficient; smaller groups can range more freely, test more prospective game, and eat more fully from kills made.

Wolf packs tend to break up in the summer. A mated pair within a large pack typically splits off in the spring to den and raise its litter. Such a mated pair may be accompanied by pups from the previous year, as well as an occasional unmated adult, while the rest of the pack continues its more free-roaming habits in smaller groups. Since in summer the diet of the wolf is more diverse than it is in winter, the summer division may also reflect the ease with which one or a few wolves can find food. A pack which has split up for litter-raising may remain divided until fall, when the pups are old enough to travel with the pack, or it may recombine earlier if the litter dies and the parents are free to travel again.

Original pack organization is based on the family group. Litter mates often stay together, either with their parents or by themselves, in more or less permanent packs; during their first year they have already worked out the social patterns which allow the pack to function as a unit. This leads to considerable inbreeding and indeed, young wolves often show mate preferences, even before sexual maturity, for members of their own litter. On the other hand, there is enough splitting off from the original litter to mix gene pools, since young wolves, especially males, are restless travelers during their first years. They may eventually leave the home territory to make contact with other individuals and small packs. As wolves reach sexual maturity they find themselves in rivalry with the dominant animals, and lesser animals frequently do not breed.

Wolves have evolved several intricate ways of maintaining group cohesion and reducing friction. One method is through voice contact. Everyone knows that wolves howl, but few people know why they howl, or that they also bark and make other domestic-dog-like noises. The howl of the wolf is too often described as "melancholy," or "blood-curdling," an indication that the human listener was as frightened by the sound as he had expected to be. But to those who know wolves well, the howl is primarily a social message, a call of greeting, of searching for contact with its own kind. Often it is a howl of sheer happiness and exuberance.

Of the many non-howling sounds wolves make, the bark may be the most startling. In the past many people, Charles Darwin among them, assumed that wolves never barked, or that if they did they must have learned it from domestic dogs. This conclusion was based on the fact that hunters and trappers never reported hearing wolves bark. The reason for this is that the bark

Gray wolf howling. *Photo: D.H. Pimlott*

is chiefly an alarm signal, is relatively soft, and is used mostly in the vicinity of the den or gathering place. Barking has now been heard from both captive wolves and wolves in the wild, and its discovery has cleared up a former mystery about domestic dogs. Why was it, people once wondered, that if dogs were descended from wolves, the one barked but the other did not? Now we know that domestic dogs come by their barking heritage honestly—and that the more frequent barking of the dog is a trait which has been intensified through domestication.

Other non-howling sounds include a squealing or squeaking noise, signaling friendly intentions, made when an animal approaches other wolves, domestic dogs, or humans with whom it wishes to make contact. Then there is the soft whine sometimes given by adult wolves, especially females, in the vicinity of young pups. This sound seems to be one of solicitude towards the pups and may also be used to call them out of the den. When wolves are excited they sometimes utter a string of emotional "wow-wow-wow" sounds, on a single tone, which sound something like those uttered by domestic dogs when they are highly excited—not a howl, not a bark, but a series of syllables which sounds to the human ear like "talking."

Though it is difficult to prove that wolves have an elaborate "language," several different howls have been distinguished. There is a call for assembling separated wolves. There is a howl of sheer joy as members of a pack gather to enjoy each other's company. Then there is the loneliness howl: a rising and falling sound with a long slide down at the end. It is often made by lonely wolves in captivity, and is heard more in the wild during the mating season than at other times. Wolves probably do not howl as they hunt, but they often do so during a rest or an assembling stop or after a long hunt when the group is scattered. This howl too is distinctive: deep, loud, and guttural, often with a few barks.

The social or group howl may also be used to call to other wolves in adjoining territories. Since wolves or wolf packs calling to each other do not then seek each other out, this howl may in effect announce and limit the territories of the howlers. Only when a stranger howls very close to a den or home site is it sought out by the resident wolves. People can hear such howls easily from a mile away, sometimes from as far as four miles.

Words do not adequately describe a sound one has never heard. But Lois Crisler gives the flavor of two wolves calling this way. "The two voices changed incessantly, rising and falling, always chording, never in unison. The chord changed in minor thirds and fifths. Sometimes there was a long note from one while the voice of the other interwove around it. The notes were hornlike and pure. The wolves would break off suddenly and there would be listening silence. We were sure they heard the voices of other wolves at the margin of sound."

The social howl is the howl of pack togetherness, the howl of pleasure and companionship. A frequent occasion for the social howl is the period of play which precedes departure for the hunt. The wolves arise from their resting spots, or gather near the den, frisking about with much tail wagging, sniffing, pawing, hugging, and pressing against each other, and then finally howling together. It is typical that during such a howl the animals move close together, apparently seeking and enjoying the physical contact. A good howling session seems to dissipate for a while the need to howl again. Observers stimulating responses with tape-recorded howls or their own imitations usually do not evoke a response for 20 or more minutes after a previous howl.

Wolf hunters have long known that wolves will respond to human imitations of their howls, and have used their knowledge to locate wolves or call them within shooting range. Imitations have been used with great success by several students of the wolf in recent research work. In wooded areas, such as northern Minnesota and southern Ontario, wolves are hard to see; but they are easy to hear and their individual and pack movements can be followed by means of induced howls. The human researcher travels about the suspected territory of the wolf, pausing often to howl (or to play a tape-recording of wolves howling)—"Where are you?"; and then to wait for the response—"Here I am."

For a hundred years people have known about "pecking order" among birds, but only with the recent emphasis on behavioral studies have we learned how mammals which live in groups use similar social devices. Patterns of dominance and submission give most of the members a specific place within the group which provides psychological and social satisfaction at the same time that it reduces tensions.

In wolves the most important psychological need seems to be to maintain contact with the animal's own group. Though it is easy to fall into anthropomorphic explanations of animal behavior, it can be said with confidence that wolves are highly affectionate animals that actively strive to create satisfactory relationships with each other. This psychological need must result from the fact that wolves hunt most efficiently in groups, but it has come to operate in the individual animal separately from hunger or sexual drives *per se*. At the same time, there is an essential biological function in the aggressive instinct of an animal to defend its territory from invasion by outsiders of the same species. Individual aggressive drives are sublimated within acceptable patterns of dominance which will not split the pack.

Among wolves, as among people, there are differing situations in which antagonisms and conflicts will occur. The intrusion of a strange wolf into the home range of a pack can lead to a severe fight in which the inferior animal is maimed or killed if it cannot escape. Within the pack itself, however, conflicts rarely go this far. A common source of antagonism is rivalry among males or females over sexual privileges. In such a case the inferior animal simply gives up its claim to the privilege also desired by the superior. This giving-up is ritualized in such a way that the superior recognizes it and is no longer stimulated to fight. The inferior wolf may give up its claim only as a means of temporary escape, though, and may attempt to claim the same privilege at some other time.

Genuine submission, unlike the above interactions, is a ritual which determines an animal's place in the pack hierarchy. Pack members usually display friendliness toward the leader, particularly before the pack leaves on a hunt. All the wolves surround the leader, pushing close to lick or tenderly grasp its muzzle, their tails wagging enthusiastically, their bodies slightly crouched, and their ears laid back along their heads. The ceremony often ends in the group howl. In contrast to such active or outgoing submission is a more passive expression of timidity or helplessness, in which the inferior animal falls to the ground, lying on its side and back and exposing its chest and abdomen. This posture of passive submission is familiar to anyone who has seen a puppy responding to the severe investigations of an adult dog.

Submission rituals are always performed in the expectation of tolerance on the part of the superior animal. If the superior wolf is actively hostile, the lesser animal has no choice but to run. But it is a tribute to the group orientation of wolves that although submissive behavior does not automatically inhibit aggression, in most cases the superior wolf does tolerate its inferior. If it were not so the wolf pack would soon be rent by the hostilities of its members, as any closed society is rent if there are no mechanisms to maintain harmony.

Hierarchy within the pack fluctuates as the animals age, and as pups, or occasionally, outsiders, become full-fledged members of the group. The original order apparently has something to do with family relations among young litter mates, since siblings grown to adulthood often continue to have a particularly close relationship, or to maintain the relative positions established during puppyhood conflicts. Sometimes young wolves may adopt the social attitudes of their socially superior parents, especially the mother, and upon maturing establish a higher position than might be expected. An adult wolf constantly attempts to widen its freedom of behavior, so that during group activities each wolf is in effect probing the reactions of its fellows—ever ready to take advantage of the weakening of any member to increase its own assertiveness. By this process, young wolves rise in the hierarchy as they mature, while old animals sink when age and disease inhibit their effectiveness.

It is common for a pack, especially a large one, to have several adult members that are gradually being relegated to the fringes, eventually to be rejected altogether. Sometimes these more or less lone wolves will be allowed to travel with, or near, the pack, feeding with it or after the stronger members have finished. At other times such animals break off, especially if the main body of the pack is openly hostile, and attempt to survive by themselves or to form a more tenuous relationship with other lone animals. Lone wolves lead a marginal existence in places where food is scarce or where the prey is almost exclusively one or more species of large animals. They are likely to be in poor physical condition, to show their anxiety when they encounter the large pack, and to have a high mortality rate. It is possible that the attitude of the pack toward its aging or injured members is influenced by the availability of prey and by the density of the wolf population; a hard-pressed pack may be more likely to reject its inferior animals.

When the leader of a pack loses its position through injury or age and falls lower in the hierarchy, the shift can be dramatic, especially during the mating season. In the early spring of 1966, researchers on Isle Royale noted that the male which had led the large pack for at least three years was limping. For several days they watched him become progressively lamer, until he was forced to lope to keep up with the walking pack. Then bad weather caused the observers to lose track of the pack for eight weeks, and when they relocated it this male was not to be seen. Two moose kills close together had been made in the intervening period and investigators discovered the remains of a wolf nearby. The carcass had been devoured and scattered, and the area showed signs of conflict; tree limbs had been bitten off and wolf hair with dried blood on it was scattered about. The few remains of the carcass showed that the animal had suffered from degeneration of the lumbar vertebrae and from arthritis in the hind legs. Apparently the leader had been confronted by one or more of the other males in the pack. Rather than retreat he had fought back, and had lost.

When the lead male or female falls from its position of dominance within a pack it does not necessarily follow that the secondary animal of the same sex assumes the leading position and so on down the line. The second animal, for instance, might have occupied its position because it was especially friendly with the former leader. More common is a rather far-reaching shuffle of the hierarchy, with a new but equally complex system of relationships developing.

The sexual behavior of wolves is a specialized part of group interaction. Mature males are usually dominant to mature females; that is, between animals which occupy approximately the same ranking within their respective sexes the male tends to be dominant to the female. Sexual activity among wolves is not random; adult wolves usually have a single preferred mate and such preferences are often formed even before sexual maturity. Wolves usually breed first at two or three years of age, but even by the end of the first year they may show signs of mate preference. Since wolves remain together in litters for the first year, and frequently longer, preferences, as we've noted, are often for litter mates.

Strong mate preferences which prevent random mating certainly have a contraceptive effect on wolf reproduction. The same effect is produced by the dominance hierarchies of each sex, which are particularly rigid during the rut. Within the wolf pack, the dominant member of each sex attempts to prevent mating by lesser members of the group. Intimidation is particularly successful if the animals are confined to a relatively small area, such as a zoo compound. Interestingly, however, although the dominant animal may challenge lower-ranking members of its sex when they attempt to copulate, not all the litters are produced by the highest-ranking animals. In fact, a lead male may be so busy discouraging other males that he himself has relatively little interest in the females.

The wolf's reproductive cycle is similar to that of other members of the genus. Pro-estrus, the period preceding ovulation when the female is sexually attractive to the male, may be as long as several weeks. Estrus itself, the only time when the female is fertile and sexually receptive, lasts three to five days. In much of the U.S. and southern Canada, most mating occurs in February, but in northern Canada and Alaska it usually takes place in the first two weeks of March. In the southern part of the Soviet Union wolves may mate as early as late December. The gestation period is approximately 63 days, which means that litters are whelped from as early as February to as late as the last part of May. Wolves produce from one to as many as 13 pups per litter, but the average number seems to be from three to five for two-year-old females and from six to eight for older animals which have bred previously. Implantation is delayed after conception, as with the domestic dog, and the fetuses grow slowly during the first part of the pregnancy so that the female is heavy and unwieldy for only a short time before whelping.

After mating, the wolf pair is constantly together and gradually begins to separate from the pack. The weeks between mating and the birth of the litter are spent looking for possible den sites, and digging out or improving the one chosen. The den may be a natural shelter such as a small cave, a crack in the rock, or an opening under tumbled roots or boulders; or it may be a burrow dug by the female herself or enlarged from that of a smaller animal such as a fox, a marmot, or a badger. Several factors determine the suitability of a den site. It should be near a source of water, since young wolves during and after weaning need water and are unable to travel great distances for it. The den must also be secluded; visibility in the immediate area is usually restricted. Successful rearing of the litter depends on the availability of sufficient large game to feed the young after weaning.

The den itself is often a rather elaborate affair. Some have several entrances, and several passages branching from the main one. These passages can be as much as 30 feet long, though most are shorter. The entrance is usually just large enough for an adult wolf to crawl into, although sometimes, several feet in from the entrance, the tunnel becomes large enough for a wolf to stand upright and even for two wolves to pass each other. The end of the tunnel is usually the place selected for the bed of the pups. This bed and the tunnel itself are very clean, since the adults come out to urinate and defecate. The female regularly cleans up the pup excreta, and licks their abdomens, a process which not only keeps the den clean, but is essential in aiding the bowel movements of the pups.

In a fascinating but claustrophobic exercise, Adolph Murie once crawled into a den to investigate the litter after the adults had left it. He found that the entrance of the burrow was 16″ high and 25″ wide. Six feet from the entrance was a right angle turn and at that point the burrow was enlarged to form a bed for the female. From the turn the burrow slanted upward for another six feet to the chamber in which the pups were bedded. The upward slant of this den had a definite function, for though melting snow had caused the female's resting place to be full of water the pups' bed was dry. This was a den that had been used by a porcupine the previous year but which had been modified by the wolves for their use.

Building a den or modifying the burrow of another animal is hard work and wolves prefer when possible to use the same den repeatedly. But hunting by man, problems caused by flooding, or contamination of the surroundings by rotting food often cause them to move, sometimes forcing them to transport the young pups several times.

New-born wolves are dark gray; small ears are plastered close to the head; eyes are tight shut. By the end of the second week, the eyes begin to open. At first they are a pale and later a deep blue, and only become brown after two months. By the third week the pups are mobile enough to stagger about and gradually to poke their noses out of the den. Wolf pups grow rapidly; one hand-raised male, when taken from his parents at four weeks of age, weighed six lbs., stood eight inches high at the shoulder, and measured 19 inches from tip of nose to base of tail. By the time he was 13 weeks old he weighed 22 lbs., stood 19 inches at the shoulder, and had a body length of 32 inches. At 16 weeks he was up to 34 lbs. At that age he was eating 2.2 lbs. of food a day.

A hand-raised wolf has the advantage of an optimum diet and regular feedings, plus protection from internal parasites. Nevertheless, even in the wild a pup with any sort of normal growth gains rapidly in weight and height and consumes a lot of food. If a litter of five pups four months old must be provided with some 11 lbs. of meat a day, it is small wonder that parents may be aided in their task by other adult wolves. It must be assumed that such cooperative raising of wolf pups has evolved because of its benefits in assuring the survival of the litter.

The birth of pups is an occasion of great interest and excitement for all adults living at the den. Wolves are conspicuous for their eagerness to take care of the young animals as soon as the mother will allow it, and adults of both sexes often vie for the chance to babysit with the pups, to regurgitate food for them, or to allow pups the most vigorous play they are capable of. Sometimes a non-parent, either male or female, will take over the almost exclusive care of the litter after it is weaned, and weanling pups can often survive the death of one or both parents. Interest in pups is not confined to mature adults; yearling wolves are equally solicitous and responsible towards babies.

By the time the pups are three to four weeks old they regularly play outside the den. Their teeth are sharp, their movements vigorous though clumsy, and they are beginning to be weaned. Since a litter of this age is still far too young to travel to a kill, food must be brought to them, and the adults bring it in their stomachs, regurgitating a supply for the young when they reach the den. Regurgitated meat is often remarkably fresh and untainted, even if it has been brought for many miles. The pups encourage regurgitation by hurrying to the adult, whining and wagging their tails, pawing and licking the adult's muzzle, and coaxing for the food they hope will be produced.

The system of bringing food in the stomach is highly efficient; it means that the adult can travel unencumbered for the many miles it may take to reach the den, and no predator can guess that the parent is returning to its pups. Although carrying food by mouth is a tiring process, wolves sometimes do this too. More often what they bring back to the den by mouth is a bone, or the relatively fleshless leg of a kill, items which serve not so much for food as for toys and exercise for the pups' eager teeth. Many other dogs, wild and domestic, regurgitate food for weanling pups, but the system is perhaps especially useful to wolves because of the distances they must travel to a kill and the quantities of food which must be provided for the pups. Incidentally, adult wolves can control their disgorge, regurgitating some of the food in their stomachs and retaining or reswallowing the

An adult gray wolf regurgitates food for a pup. *Photo: D.H. Pimlott*

rest for themselves. So strong is the urge to disgorge food for pups that wolves in captivity will often regurgitate for the benefit of domestic puppies that live with them.

Being bound to a single site is difficult for wolves, accustomed as they are to making a kill and remaining nearby until it is consumed. And so when the pups are old enough to travel, usually some time between six weeks and two months, the family leaves the den and begins to move about more freely. The young are still not able to travel for many miles at a time, so for the next few months they use resting sites. Such spots are located along travel routes, and have some of the characteristics of the den site. They are near water, in secluded areas, and in places where a mixture of vegetation allows for both sheltered beds and open play areas. Here the pups play and sleep for several days at a time while the adults hunt the surrounding country. The resting sites also serve as assembling points for the adults; the parents are found there when they are not actively hunting. Often there is a string of such resting places along the hunting routes, and the family uses them alternately throughout the summer.

In following the pack, the pups gradually become less dependent. Young wolves do not automatically know how to hunt; it is a process they must learn through imitation and trial-and-error. They practice stalking and pouncing movements of adults as soon as they can walk, engaging in mock attacks on each other and attacks in earnest on the insects, birds, and small mammals around the den. As they grow they perfect these movements but initially show great caution in approaching larger prey. Wolves are naturally very suspicious, even timid, of new elements in their surroundings and a pup first faced with a caribou or moose keeps a respectful distance indeed. The young wolf's first hunting success is likely to be against particularly weak prey and only gradually does it become bold and experienced enough to be of real assistance in group hunting.

Prenatal and juvenile mortality is high among wolves. On Isle Royale, the packs do not reproduce every year, but even when a litter is born its chances of reaching adulthood appear to depend on how well the moose are thriving. For pups, the period spent at the den is critical. Young wolves are susceptible to diseases and parasites; they also fall victim to accidents or to predators, such as bears, which are sometimes found around dens robbing cached or regurgitated food. Even among domestic dogs, approximately one third of the pups are lost before weaning, and the percentage must be as high among wolves. Pups have to eat regularly, and after weaning, starvation accounts for the loss of many. Despite the many mortality factors weighing on the young, wolf populations are able to increase gradually where prey is abundant and hunting has been eliminated.

Though wolves share their ranges with other predators, relations between them seem generally to be neutral. Large cats can be an exception: in the Far East, the tiger is a well-known killer of wolves, and wolves have increased noticeably in places where tigers have disappeared. Throughout much of the northern hemisphere bears are found where wolves are. The two species are not competitors, since bears do not hunt large hoofed animals systematically, but they do forage around wolf dens for meat scraps. When the bear is a large one, like a grizzly, it is more than a match for a single wolf and only a pack, working together, can hope to drive it away. This they do by repeated lunges and harassment, but they must take constant care to avoid the bear's arms; a wolf caught by a bear would be instantly clawed, bitten, and then crushed to death. Wolverines also forage at wolf dens, and an account of a wolf driving off a wolverine suggests that the former has a great respect for the other predator's strong arms and sharp claws. Wolves may be the most important natural enemies of the lynx; in Finland, in the last century, when the wolf population was sharply reduced, lynxes became much more numerous for a time. When coyotes and wolves occupy the same area, there is no question but that wolves sometimes kill their smaller relatives and it is possible that the presence of many wolves within a territory may drive coyotes away.

On the other hand, foxes and wolves share the same range. Though instances have been reported of wolves killing foxes, foxes do feed on wolf kills when the wolves have withdrawn. Nor do they seem always to fear wolves, particularly when they are able to dodge into dense cover to protect themselves from attack. Wolves in turn use fox burrows as dens, and also scavenge fox dens when food is scarce.

A curious relationship occurs between wolves and ravens. In both Alaska and Isle Royale, ravens follow hunting wolf packs to feed on the kills. They also pick apart wolf droppings and consume whatever is edible. A flock of ravens will fly ahead of a traveling pack, perch in trees until the wolves have passed, then fly forward again to catch up with them. When the wolves rest, the ravens delight in playing with them, diving down or even waddling up on foot to peck at the predator. The wolf will leap to its feet and stalk the bird, which waits until the animal is within a foot or two before flying a short distance away to await another stalk. When the wolves tire of this game of tag, the raven may squawk to try to keep it going. Though the dogs often seem within an inch of snapping up a bird, in the wolf dung examined on Isle Royale, raven remains have never been found. Either ravens know exactly how to elude them or else the wolves are not really trying.

Progenitor of the domestic dog, the wolf still comes in contact with man's favorite pet both as friend and as rival. The close relationship between the two species leads captive wolves, and often wild ones as well, to act with friendliness towards dogs. People who have raised wolves have noted that wolf pups court and become attached to adult dogs just as they would to adults of their own species, and that adult wolves respond to dog pups as they would to their own. Even among adults of the two species, relations are often amicable; captive wolves will play with dogs, are willing to accept them as part of a composite grouping which substitutes for the wild wolf pack, and of course will mate with dogs. In the wild, although there are many accounts of wolves killing and eating domestic dogs, especially in Europe, the accounts of sexual and social contacts are also numerous.

There may be some differences in the way wolves respond to various breeds of domestic dogs. An experiment in which captive wolf pups were introduced to dogs of two different breeds—German Shepherds and Samoyeds—showed that the wolves acted towards German Shepherds of all ages as though they were wolves. They were submissive and friendly towards animals older than themselves and initially hostile to younger pups, as they would be to younger members of their own species. On the other hand, wolf pups reacted with instant hostility to adult Samoyeds, snarling, biting, and springing at them. There is something about Samoyeds, either their posture, their color (white), their tightly curled and unwolf-like tails, or their scent, which seems strange to wolves, whereas German Shepherds, which bear a close external resemblance to wolves, seem familiar. German Shepherds are certainly genetically at least as distant from wolves as are Samoyeds, so here behavior and external appearance were evidently all-important.

The extent to which dogs and wolves have bred in the wild is subject to dispute, though the fact that crossing results in fertile hybrids is borne out by well documented cases. A famous wolf-dog cross occurred in Colorado in 1923. A female wolf named Three Toes lost her mate of many years and began to visit a collie owned by a local rancher. When the collie took to leaving the ranch to run with her, the owners locked him up at night. Three Toes maintained her contact by howling, and finally one night helped him to dig free. Several months later, Three Toes and her litter of five collie-wolf pups were captured. One of the female pups, which clearly showed collie ancestry in her reddish color and slender build, was raised and later bred with an airedale to start a strain of cross-bred animals which continued for three generations. Some of these dog-wolf descendants were notable for their intelligence and affectionate disposition.

Among sled animals of the north there have been numerous half- or quarter-wolves which have made outstanding work dogs. But much of the testimony about wolf-dog crosses rests on the possibility of such matings rather than on observed and proven instances. Early travelers in North America contended that Indian and Eskimo dogs were obviously wolf-dog mixtures and that the natives deliberately tied out their bitches in heat in order to obtain extra-strong and extra-intelligent wolf-dog draft animals. Such contentions are not necessarily reliable: unless a large, apparently wolf-like dog has an obviously distinctive feature, such as a curled tail or a spotted pattern, the two species can sometimes only be distinguished reliably by examining the skull. At least one student of American aboriginal dogs contends that they show little evidence of extensive crossing with wolves. It is probable that more wolf-dog crosses are claimed than actually occur.

As suggested earlier, perhaps no other predator has been more systematically hated, feared and hunted by man. The origin of this feeling may be in the superstitious dread felt by primitive people towards animals stronger than they. But civilized man's antipathy towards the wolf grows from its role as livestock killer. Under natural circumstances, wolves prey on the wild animals that make up its natural diet. But natural conditions no longer prevail over much of the wolf's range; in some places they have not done so for thousands of years. As man the hunter, herder, and agriculturalist expanded throughout Europe and Asia, eliminating wild game and putting domestic stock in its place, the wolf turned to this stock for food. (In America, pre-European Indians, since they were not herders of livestock, were in little conflict with the wolf.)

Domestic livestock is singularly defenseless against the depredations of wolves. Smaller animals like sheep and goats have no natural defenses at all, and when they are turned out without the protection of a shepherd are ridiculously easy to kill. A settler in Oregon in the last century, for instance, lost seven sheep to one wolf within the fifteen minutes it took him to go into his house for a gun. Larger stock may fare better; canny wild horses on the American plains were known to make a stand against wolves, striking out with their hoofs as they stayed bunched together, a defense which could discourage even a large pack. But strayed domestic horses would turn and flee, exciting the wolves to pursue and run them down. Cattle, especially the large, stolid, de-horned animals which by the end of the 19th century made up the vast majority of herds on the American plains, fared poorly when attacked. Unguarded herds of domestic reindeer also fall victims to wolves, which kill fawns—sometimes adults too—and scatter the herds.

In western Canada, at the end of the last century, it was accepted that a full-grown wolf could account for a thousand dollars worth of damage to livestock a year. In 1904, around Meeker, Colorado, it was claimed that wolves were taking about one-fourth of the calf crop per year. Not only do wolves kill, they can severely maim surviving animals. They attack domestic stock much as they do wild prey, biting at the flanks to bring down the animal. Ranchers often found cattle with their tails severed by such attacks, or a living animal with huge chunks of flesh torn from its hindquarters.

It was wrongly concluded in the past that wolves must select natural prey as they do livestock. The argument was that wolves could not be picking on the weak or disabled among wild populations when they so frequently killed large numbers of the fattest, most vigorous young adults among cattle herds. Such arguments ignored the fact that eons of natural selection have provided wild animals with their own defenses, and that millenia of selection in the opposite direction have deprived domestic stock of exactly those same defenses. For man's benefit, a Hereford cow has been bred to lose its swiftness, aggressiveness, wariness, and often its horns. It is hardly surprising that such losses work to the advantage of the wolf.

Tales of wolves attacking man are as old as the history of Europe and Asia. Many such attacks are not unprovoked; they may be the attempts of desperate animals, held at bay by man and his dogs, to break loose from their encirclement. The vast majority, though, are attacks by rabid wolves, which do not distinguish man from any other object that attracts their attention. The stories of unprovoked wolves in Eurasia killing and eating people are so numerous, however, that a few authorities on wolf behavior give credence to at least some of the cases.

It is quite different in North America. Here, stories of wolf attack are few, and despite several exhaustive studies, not a single such story has been proved. In most versions, a lone man has found himself surrounded by wolves which he assumes are about to attack him. When he escapes with his life, his story quite naturally becomes one of unprovoked attack. Wolves will indeed follow men on foot or horseback, will sometimes gather around a person apparently out of curiosity, and in times of famine will attempt to kill the horses or cattle penned or tied nearby. But neither the U.S. Biological Survey (which made a 25-year study of the question) nor any other unbiased investigator has ever discovered a verified instance of unprovoked wolf attack on man in North America.

One immediately wonders why the wolves of Eurasia should be different. One inescapable conclusion is that the wolves there, which have been far less carefully studied in the wild than those of the Western Hemisphere, are certainly not guilty of all the charges against them. Whenever stories of attacks by predators on man are carefully investigated, the great majority turn out to be fabrications, innocent or otherwise.

But how can one explain those attacks which may be genuine? A complete answer must await thorough study: it could be that the long presence of man in Europe and Asia, and the consequent alteration of the habitat, have caused some changes in the social behavior of wolves and their basic feeling of respect for human beings. For instance, in places where human population is dense and poverty severe, famines and epidemics result in large numbers of human corpses which are fed on by wolves. Such disasters are known to lead to the habit of man-killing among the large cats; perhaps in some instances they could account also for wolf attacks. In places like India, where wolves are reputed to carry off children, it may be that the disappearance of natural prey and the ensuing famine among predators, plus the vulnerability of children in isolated villages, may combine to account for some such instances. But in view of the totally negative evidence of unprovoked man-killing by wolves in the one area where the problem has been studied, all reports of wolves killing men must be viewed with great skepticism.

Obviously wolves do not regard human beings as prey. First of all, they learn to hunt basically by experience, not by instinct, and they never have the opportunity to learn to hunt men. Secondly, men are predators themselves. They stalk their way through woods and across streams, quietly and deliberately, as wolves do. Many human behavior patterns may indicate to wolves that man is the hunter, not the hunted, the predator, not the prey.

Interestingly, man's hatred of the wolf is balanced by his traditional myths of wolf-children—human infants supposedly adopted and reared by wolf parents. Such myths originated long before Romulus and Remus and have survived in the recent case of the so-called "wolf-boy" of Lucknow, India. These stories are based only on children having been seen near wolves, or, as with the Lucknow tale, on the fact that an abandoned, feeble-minded child crawled on hands and knees, lapped water from a dish on the floor, and was unable to utter words—traits which were then imagined to have resulted from continued contact with wolves. But since wolves nurse their pups for less than two months and subsequently feed them regurgitated food, it would be impossible for a human infant to survive this way. There are no authenticated cases of children either being reared by wolves or indeed having any close contact with them at all.

Over the centuries, man has devised a wide variety of wolf-traps and hunting methods. Primitive devices included the pit trap: a hole dug along a wolf runway and so disguised by a layer of dirt or vegetation that the traveling wolf tumbled in, to be pierced by sharpened stakes at the bottom or trapped by wooden or stone sides. Sometimes the trap was baited and the animal fell in while trying to reach the meat. A particularly effective and diabolical weapon, the piercer, has been used by Eskimos and other northern people. The piercer is a sharp object, most often of whale baleen, which is twisted or coiled, tied with sinew, wrapped in fat or blubber, frozen and then untied. The bait is scattered about liberally, the wolf bolts some of it, and as the fat thaws in the animal's stomach the piercer springs open to perforate the stomach wall and cause a very painful death. A smaller variation of the piercer, using a gull's wing, is employed to kill foxes.

For hundreds of years, hounds were bred in Europe and Asia especially for wolf hunting, and two breeds, the Irish wolfhound and the Russian borzoi, were particularly successful. Such a hound must be both swift enough to overtake a wolf and strong enough either to kill it or, in the case of the lighter-weight Borzoi, to hold it until hunters arrive. One efficient method of wolf destruction is the den hunt. If the man finding the den cannot crawl into the opening to reach the pups he either sends in a terrier strong enough to dispatch them or pull them out, or he reaches in with a hooked stick or wire to twist into the pup's fur and drag it out.

The coming of the gun made a great difference to wolf hunts. Hunters often banded together to drive an area, encircle the wolves, and shoot them as they attempted to break out. A good individual stalker might also be successful, sometimes enveloping himself in a wolf pelt to crawl within shooting range. A set gun—cocked, lashed in place, and triggered by a trip cord—was also used, though such a device is extremely dangerous to man, domestic stock, and other wild animals which could inadvertently trigger it. But the steel trap and the widespread use of poison have accounted for the largest numbers of wolves, at least in North America.

Successful trapping depends on a thorough knowledge of the routes and resting sites a wolf is likely to use. The trap is buried cautiously so that no human scent adheres to it or is allowed to reach the ground around the site. The bait is either meat or, better, a composition of wolf urine and ground anal glands, rotted meat or fish, with glycerine as a preservative. The wolf is attracted to the scent, pauses to investigate, and is caught when it steps onto the trap. The hunter following his trap lines comes on the caught animal and shoots or clubs it to death. The reactions of a wolf caught in a trap vary. The frequency with which the animal can pull, twist, or gnaw itself free is attested to by the number of peg-legged or otherwise crippled wolves which ranged the west during the heyday of the wolf trapper. Yet sometimes the trapped wolf simply lies down to await the trapper and offers no resistance when discovered. This docility has allowed researchers in Ontario to use a trap with one spring removed that does not injure the animal; the wolves so captured are tagged or fitted with radio-equipped collars for further study.

By the middle of the 19th century, strychnine was widely and cheaply available in the American west for the destruction of wolves. The motive then for wolf killing was not primarily the protection of livestock, for cattlemen were still few and far between. Wolves were poisoned for their pelts, which found a ready market for use in overcoats for the Russian Army. The professional wolfer followed the wolves which followed the bison herds, and between 1860 and 1885 the number of wolves killed was immense. During the winter of 1861-1862 three wolfers in what is now southwest Kansas killed more than 3,000 wolves, coyotes, and kit foxes. Wolf-poisoners endangered every meat-eater within range, and coyotes, kit, gray and red foxes, skunks, and scavenging birds died in great numbers. In fact, it was the activities of the wolf-poisoner that virtually wiped out the totally harmless and highly beneficial kit fox, a small creature which was often first in line to feed on poisoned carcasses.

By the latter part of the century the interests of the stockman had taken precedence over the profits of the pelt hunter. Stockmen's groups as well as local governments offered bounties on wolves, and every stockman thought it was in his interest to carry poison on him at all times and systematically poison any animal carcass he came upon. The numbers of wolves killed after about 1885 were far smaller than those of earlier years, for there were simply fewer wolves about and less open range for them to roam.

As the total number of wolves declined in the American west, more and more attention was paid to the capture of individual wolves whose depredations on livestock were especially serious. These wolves, called renegades, were the exceptionally canny individuals that had survived the usual hunting methods, often had injuries which made them dependent on livestock rather than wild prey, and because of their experience were the most difficult to capture. Renegades learned never to return to a kill, always to investigate any scent or bait with extreme caution, always to kill far from the den, never to show themselves in the daytime, to move their pups at the slightest indication of detection, and even to abandon a litter if necessary to save themselves. Their physical deformities, the areas which they haunted, or their habits led to names like Three Toes, Peg Leg, Old Lefty, the Phantom Wolf, the Custer Wolf—all animals which at some time or other became the locally famous objects of intensive hunting.

To be successful at capturing a renegade, the hunter had to be unusually knowledgeable of his individual quarry; often he spent months on the track of such an animal. The Custer wolf roamed the Black Hills around Custer, South Dakota, for six or seven years despite a $500 bounty for its scalp—and during that time is reported to have destroyed $25,000 worth of livestock. The government hunter who finally killed the animal in 1920 worked constantly on that job alone for seven months. An equally famous renegade was Lobo, a 150-pound male that ranged the Currumpaw region of northern New Mexico from 1889 to 1894 and caused great damage to cattle and sheep before Ernest Thompson Seton captured and later wrote about it. Three Toes, of Harding County, South Dakota, eluded at least 150 men for 13 years before being taken by a Biological Survey trapper. Stockmen contended that Three Toes had caused losses of more than $50,000 during its life.

Accounts suggest that a great many of the famous renegades were livestock killers because of injuries suffered at the hands of man. Certainly the renegade killed more often and more destructively than a normal wolf; because it was forced to move on, it never returned to a kill, nor took up with a pack which could help provide it a normal diet of wild animals. Hunts for renegades show us the wolf at its canniest, for the intelligence necessary to survive concerted hunting over a number of years was considerable. But these were abnormal wolves, in which all other instincts were subordinated to that of escaping the clutches of man.

One of the oldest devices of wolf control is the bounty system. In existence in ancient Greece, it was widespread in England from Norman times until wolves disappeared from the British Isles, and in the United States developed a great volume of legislation. Bounties were offered by state or territorial governments, counties, and townships, and often by stockmen's associations as well. In colonial days the sum of three or four dollars per wolf was average, while in the 19th-century West as much as $30 to $40 was paid per animal, a considerable sum in those days. Bounties are still paid on wolves in parts of Canada and Mexico.

Bounties may have helped account for the decline of wolves; certainly they spurred on determined hunters. Legally, the chief trouble with the bounty system is that it is open to flagrant abuses. Various types of fraud were commonly practiced throughout the west, such as presenting the scalp and ears of coyotes or domestic dogs as those of wolves, taking scalps across state or international boundaries to profit by higher bounties on the other side, and even releasing trapped females so that they might continue to reproduce and support the bounty-hunter. When control of any predator is necessary, a far better system is the employment of trained and salaried hunters and trappers to carry out planned control measures.

Fortunately not all men hate or hunt wolves. When man takes the trouble to become friends, a close relationship can develop. Wolf puppies taken from a den at the age of three or four weeks—the time at which they begin to form social bonds with their litter mates—can become very attached to the people who care for them. Wolves raised by man transfer their normal social attachments to people and their domestic dogs, and can display deep affection and trust for their human friends. Some people who have raised young wolves, however, have encountered difficulties in handling the animals after they become adults. It is not like raising domestic dogs; although a wolf pup may be controlled for a time when it is young and small by the same sort of strict discipline that will work with dogs, an adult wolf will not accept stern control, and attempts to discipline it by blows or physical control lead only to aggressive responses. Domestic dogs have lost many aggressive tendencies through long and selective breeding. But if a wolf thinks itself threatened unreasonably it will fight back. It must be persuaded with infinite patience, in a relaxed atmosphere, to maintain its friendly relations with people.

Accounts of wolf-taming should not encourage anyone to go

out looking for one to raise as a pet. But they do reinforce observations made in the wild of the central importance of social contacts in the life of the wolf—and of the sophisticated techniques they have evolved for successful group living. Certainly it is this knowledge of the wolf as a social animal, as much as information on its ecological role as a predator, which has stimulated today's increasing interest in every aspect of wolf life. It is suggested that a sound knowledge of the way wolves handle their social organization and sublimate their aggressions to the benefit of the pack will throw some light on how the social organization of other predator animals, including man, has evolved. Whatever social benefit people may derive from the study of wolves, the fact that we now know enough about this remarkable animal to appreciate it is surely ample reward. Perhaps, if our appreciation is sufficiently keen and widely felt, we may yet find the means to guarantee the wolf its existence in the wild. And if wild wolves live on, man will surely be the richer.

THE RED WOLF (*Canis rufus*)

The United States is the home of the world's most critically endangered wild dog. This is the red wolf, of which perhaps no more than 100 pure individuals are left. Only within the last ten years has its plight come to the attention of the American scientific community and the public at large; now, as often happens, we run the grave risk of losing an animal which we have only just begun to know.

The red wolf looks like a very large, long-legged, tawny coyote. Its head and body are from 42 to 49 inches long, with the tail another 13 to 17 inches. It stands from 24 to 30 inches at the shoulder and weighs between 50 and 80 pounds. The red wolf often howls like a gray one, but may also bark and yelp, and two or more red wolves calling to each other sound very much like coyotes. In its present range the red wolf's diet consists primarily of rabbits, muskrats, nutria and an occasional waterfowl. Formerly it preyed on wild turkeys, and deer no doubt

After Hall and Kelson (1959)

made up a larger part of the diet than they do now. Red wolves are capable of killing calves and disabling adult cattle, but the majority of large animals eaten are probably found as carrion.

Red wolves associate in pairs, in small family groups, or in packs of up to a dozen. Litters range from three to 12 pups, at an average of half a dozen, but pup mortality is very high, perhaps in part because of the prevalence of hookworm in the hot, moist terrain to which they are now confined. Wolf pups mature at the end of their second year but as adults frequently remain with their parents to form extended family groups.

Red wolves once ranged from Florida to Texas and from the Gulf of Mexico up the Mississippi Valley as far north as southern Missouri and Indiana. Within this original range, eastern animals were especially large, with size decreasing in the western part of the range. The animal also originally occurred in two color phases, one the tawny or cinnamon-buff which has given it its name, the other a dark gray or black. The black phase was common in Florida and other eastern portions of the range, while tawny predominated in the west.

Red wolf of typical tawny color. The red wolf, found only in the southern United States, is in grave danger of extinction. *Photo: Dr. M.W. Fox*

Red wolf of the now-extinct black phase, photographed in Madison Parish, Louisiana, in 1935. *Photo: Tappan Gregory, Chicago Academy of Sciences*

The red wolf, including all black-phase animals, has long since disappeared from the eastern part of its range. Large, reddish animals are found throughout eastern Oklahoma and Texas, and in parts of Arkansas and Louisiana. Until recently their numbers indicated that the red wolf thrived west of the Mississippi but about ten years ago it was pointed out that these so-called red wolves showed a wide variety of size and skeletal characteristics —ranging from those of coyotes to those of the original red wolf—and the possibility was raised that in the western part of its original range the red wolf had hybridized with the coyote.

This suggestion of extensive hybridization between two species presumed to be distinct intrigued zoologists; it could mean that what were thought to be red wolves might in most cases be hybrids, and that the so-called red wolf might not in fact be a separate species. Recent investigations have established that the red wolf probably *is* a distinct species, that in the western part of its original range it has hybridized with the coyote; and that the pure red wolf is in imminent danger of extinction.

An extensive study of the skulls of gray wolves, red wolves, coyotes, and red wolf-coyote hybrids has shown that the skulls of red wolves differ from those of the gray in several proportions, including a flatter forehead, a longer, more slender muzzle and canine teeth, and less widely flaring arches in the cheekbone. In all these proportions red wolves closely resemble coyotes. The diagnostic distinction between skulls of those two is one of size. Within each sex there is no overlap in size between the skulls of the smallest red wolf and the largest coyote.

Hybridization between red wolves and coyotes is now well confirmed. Skulls collected from east Texas since 1962 run the gamut in size from coyote to red wolf with every intermediate size well represented. Farther west, on the Edwards Plateau of Texas, where red wolves once occurred, only coyotes are now found. Eastward, in Arkansas and Louisiana, where coyotes were not originally found, coyote-sized animals are beginning to appear. At about the turn of the century, changes in the habitat of the Edwards Plateau, brought about by intensive farming, ranching, and settlement, coupled with the rigorous poisoning campaigns against wild dogs, caused a decline in the populations of both red wolves and coyotes, especially the former.

This decline caused a breakdown in the isolating mechanisms between coyote and red wolf; mixed mating occurred and the result was a hybrid swarm. The swarm gradually moved into east Texas, entering areas where the pure red wolf had disappeared, and eventually populated parts of Louisiana. On the Edwards Plateau, only pure coyotes remained. The hybrids are continually being reinforced, from the west, by coyote genes. While there is little red wolf reinforcement, enough of those genetic factors still occur in the swarm so that throughout east Texas large, red, wolf-like animals continue to crop up.

Although pure red wolves seem to be extinct in Arkansas, Oklahoma, and northeastern Louisiana, areas where they were still fairly common in the 1930's, a pure population remains in the coastal plain of Texas, from Brazoria County south of Galveston northeast to Orange County. Possibly red wolves also occur in adjacent coastal areas of Louisiana. The habitat of this part of Texas has been changed relatively little by man's activities, and extensive predator control programs have never been carried out here. Further, the hybrid swarm of red wolf-coyotes from farther inland has not penetrated the coastal area completely.

Several active steps have been taken to preserve the red wolf. The first was to place it on the U.S. Bureau of Sport Fisheries and Wildlife's list of endangered species. Next, the Bureau's Division of Wildlife Services began a study of the Texas red wolf population and is trying to convince local stockmen to cooperate in its preservation. Rather than responding to occasional red wolf depredations on livestock with widespread poisoning, the Bureau now tries to live-trap individual offenders.

The number of red wolves today is so small that it is imperative to establish a captive breeding population. Animals live-trapped by the Division of Wildlife Services are turned over to the Red Wolf Committee of the Wild Animal Propagation Trust, which has formed breeding groups at several zoos around the country, notably the Tacoma Park Zoo in Tacoma, Washington. The Red Wolf Committee has also sponsored an investigation of the biochemistry of the red wolf in hopes of finding a simple method of distinguishing pure red wolves in the field.

Although breeding red wolves in captivity should be relatively simple, maintaining the wild population and keeping it free from coyote contamination is another matter. The concern of the U.S. government and the support of international groups like the World Wildlife Fund and the International Union for the Conservation of Nature guarantee, one hopes, that all reasonable measures will be taken. Just possibly, the red wolf still has a chance to survive.

THE COYOTE (*Canis latrans*)

Also called the brush cr prairie wolf, the North American coyote is a fabulous animal, if only in the sense that it has been the subject of innumerable tales about its cunning, its behavioral peculiarities, its raiding of livestock, and its supposed magical qualities. Even its name is exotic, for "coyote" is the Spanish adaptation of the Aztec name *coyotl.* By the early 19th century the Spanish term was being used by Americans, though the spelling–which in old texts is rendered *ciote, cuyota, kiyot,* or *cayeute*–took a long time to settle down.

The behavior ascribed to the coyote is often more fantasy than fact yet, like the fox and the jackal, this versatile and almost indestructible animal has brought itself to the attention of man by its sheer skill at survival. In a family of generalists the coyote is the most adaptable member–it seems, sometimes, that a coyote can live anywhere, eat anything, escape any hunter, and reproduce itself under almost any circumstances. It has even withstood the transformation of its habitat by man and at the same time expanded its range.

For the most part the coyote was, and is, an animal of the American west. The white man originally encountered it at the point where the forests gave way to prairie west of the Mississippi River. In this western range the coyote was found from as far north as Alaska south through Mexico to Costa Rica, and over much of the area it was conspicuously common. Now it is nowhere as prevalent as it once was–thanks to poisoning by man and habitat changes–but it is even more widespread. Spreading steadily northeastward, coyotes now live in Quebec, are well established in the Adirondacks of upper New York State, and even occur in Maine. They are still not found in significant numbers in the southern Atlantic states.

There are several apparent reasons for this extension of range. In the northeastern U.S. and eastern Canada, man has cleared dense forests, producing broken patches of woodland, scrub, and farmland attractive to coyotes. He has eliminated the gray wolf from most of the area, thus removing a competitor and occasional predator of the smaller dog. Perhaps, too, the pressure of poisoning and shooting in the west has encouraged the coyote to move east where, though not welcome, it is not so subject to constant and highly organized harassment.

The quickest way of describing a coyote is to say that it looks like a small wolf. In fact, distinguishing between the two can be a problem when they both occur in the same range. But each is subject to such great individual variation that only by describing the characteristics of both species as a whole do the differences become clear. The coyote's length of head and body ranges from about 32 to 37 inches, with a tail that may be as long as 16 inches. At the shoulders the coyote may stand as high as 26 inches, though several inches less is average. Weight varies greatly, from about 20 to around 50 lbs., the females being lighter and smaller than the males. A coyote's ears are pointed, as is the nose, and the nose pad is narrow. The tail is rather bushy and frequently hangs down as the animal runs. It is buff or grayish on the back, neck, and head, with light or whitish underparts and rust-colored legs, feet, and ears. The tip of the tail is dark or black. There is considerable color variation among individuals, though not as wide as that of gray wolves.

Although the similarity between small wolves and large coyotes can lead to confusion, researchers have discovered that in two animals of the same size, the coyote's skull proportions are distinctive, with a relatively large brain case, slender muzzle, and long tooth row. (This method of analysis is, of course, hardly useful in the field, since it requires a dead subject.)

As interesting as the coyote's physical structure may be, it is his behavior, especially his famous song, which has attracted the most public attention. Though it does not form large packs as wolves do, the coyote barks, yaps, or howls mainly as a means of

American coyote in thick winter coat. *Photo: Robert R. Wright*

social contact. It is this social instinct which is served, especially at dusk, as a lone animal or a hunting pair sets up a howl that rises and falls in a cascade of sustained notes interspersed with barks. Other coyotes within hearing may then take up the chorus. As Archer Gilfillian describes it, "The peculiar thing about the coyote song is that when two coyotes are singing a duet, as they are very fond of doing, they do not bark haphazardly or in unison, but they catch each other up with lightning-like quickness. . . . On a cold winter's night you will hear this two-piece orchestra tuning up on some distant butte, and then miles away in another direction you will hear an answering chorus; then the cry will be picked up in still other directions, until it seems as if the whole landscape were tossing this weird melody up towards the cold and unappreciative stars." Small wonder that the scientific name of the coyote, *Canis latrans,* means "barking dog."

The coyote prefers an open or partially wooded habitat to dense forest, and is perfectly at home where there are few trees—in plains and deserts. Within this general habitat it may be found in low, swampy areas like the lower Mississippi basin, in

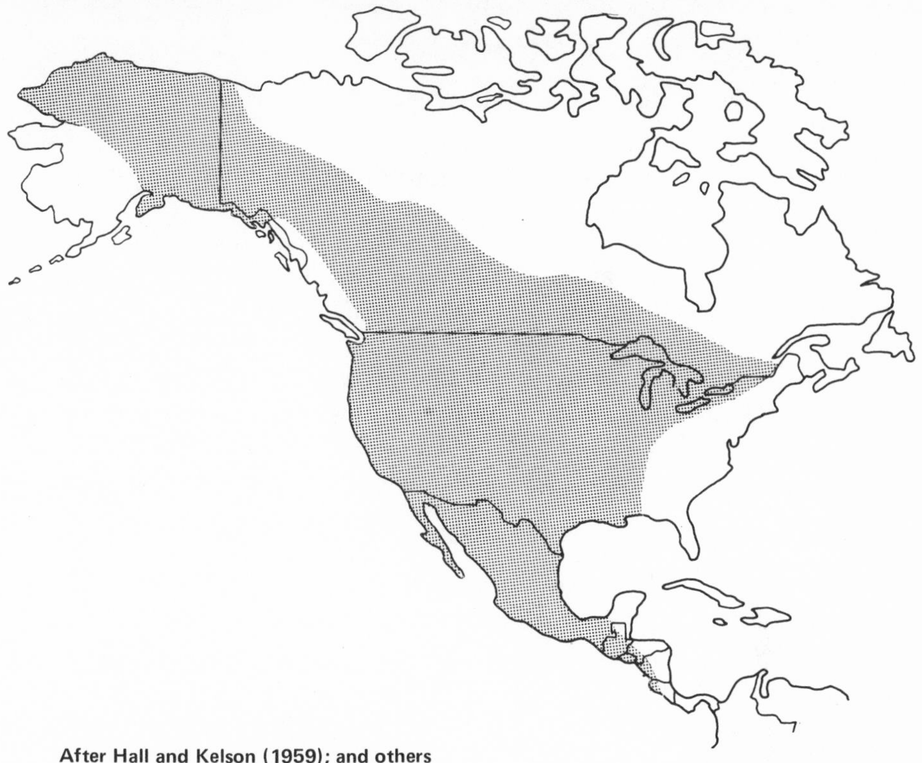

After Hall and Kelson (1959); and others

the mixed hardwood and coniferous forests interspersed with farmland of the northcentral states, on the high parks and open mountain slopes of the Rockies, on the treeless central plains of the U.S. and Canada, and in the semi-deserts of the southwestern United States and Mexico. Since its diet consists primarily of small mammals, it is not tied to the ranges of the large herbivores. It is tolerant of both the extreme cold and great heat.

Coyotes are territorial: adults prefer a hunting range which they leave only under compulsion and to which they return by choice. They will, of course, stray beyond their territories if they have to—in search of food or a mate—and young animals may range widely before selecting and settling down in a certain area. A coyote's territory is much smaller than that of the wolf: in a region well stocked with prey it is perhaps ten miles across; in a poorer area it may be as much as 25 miles. Many animals tagged, released and subsequently recaptured will be found where they were first trapped—though one tagged animal traveled 125 miles and crossed four mountain ranges in two weeks. Coyote hunting territories are not exclusive; easily available food, such as a supply of carrion, will bring many animals together. Within the more confined denning areas, however, coyotes are aggressive towards strangers.

Like many wild dogs, the coyote was originally diurnal and nocturnal, hunting at dawn and dusk, and being seen during the day as well as heard at night. Only in areas such as the southern deserts, where high daytime temperatures are debilitating, have activities been restricted by natural conditions. Now, of course, no coyote which values its life regularly ranges abroad in the daytime in places where every man's hand is against it, so that over large portions of its range the animal is now effectively nocturnal.

Coyotes probably have as varied a diet as any carnivore. They eat rabbits, rodents, birds, amphibians, reptiles, and fish; carrion of all kinds, including each other; all sorts of vegetable matter; occasionally they even eat such tidbits as tanned leather. A comprehensive study of the food habits of the coyote, made by Charles Sperry as the result of an analysis of the stomach contents of over 8,000 coyotes, showed that rabbits and hares are most important. They occurred in coyote stomachs 43 percent of the time; and are especially crucial since young rabbits are an essential part of the diet for coyote pups being weaned. Of second importance is carrion, mostly of horses, cows, and sheep. Third place in the coyote bill of fare is taken by rodents—field mice especially. One stomach contained 20 almost whole mice, as well as a song sparrow. The remainder of the diet is filled by domestic livestock, large game animals, birds, non-mammalian vertebrates, insects, and vegetable matter. Sperry mentions a professional bird-catcher from New Mexico whose stomach contained a robin, a bluebird, a meadowlark, and a blackbird. Then there was a Texas coyote with a stomach full of rattlesnake, another from the same state which had eaten a gopher turtle, and an insect-lover from Arizona with a stomach-load of 500 grasshoppers.

The coyote's reputation as a predator of large game and livestock is responsible for the vehemence with which it is attacked through much of its range. In Sperry's study most of the remains of large game found in coyote stomachs was deer, with a little antelope and traces of bear, bison, elk, and bighorn sheep. Deer formed 3.5 percent of the food in bulk and occurred in 6 percent of the stomachs. A good portion of the deer is certainly carrion, as evidenced by the fact that there is a considerable rise in deer consumption on the part of coyotes during and immediately following the open season on deer in several states. Some also is "carrion on the hoof," animals wounded by hunters and later killed by coyotes. In addition to increased deer consumption during the hunting season, a peak occurs late in the winter, when severe weather may debilitate or trap deer, and again at fawning season. Despite this, coyotes frequently are seen near deer with neither species paying any attention to the other.

So large game, partly as carrion, forms a significant but small part of the coyote diet. But are coyotes the great livestock killers they are reputed to be? Sperry found livestock in only about 20 percent of the coyotes examined and calculated that it made up about 13.5 percent of the annual coyote diet in terms of bulk. Nearly all these remains were sheep and goats; very few were calves, colts, or pigs. Coyotes are indeed capable of killing calves, but do so rarely, and some cattlemen consider the coyote an asset because it kills so many rabbits and rodents.

Sheepmen, however, feel different. Unquestionably coyotes do kill sheep, in some areas both frequently and in large numbers. Not all coyotes prey on sheep, however; the sheep-killing coyote is often an old or crippled animal, sometimes one crippled in a trap. In fact trapping, particularly by novices, may actually increase the amount of sheep-killing by coyotes, since it increases the numbers of peg-legged coyotes which have escaped from traps. Maimed coyotes eat about 50 percent more livestock than normal animals, and take a third fewer rabbits and rodents. Some predation on sheep is seasonal; the high seems to occur in the spring and summer, which coincides with lambing time for sheep, and denning and family-raising time for coyotes. In sum, the sheepman has a genuine grievance against coyotes, especially against certain individual animals.

The coyote typically hunts haphazardly to pick up its miscellaneous prey. During the mating season and while the family is being raised, coyotes may hunt in pairs or in family groups. However, since the small animals they depend on do not require true pack hunting, coyotes usually hunt more or less independently of one another.

An exception is the group tactic used when coyotes attempt

to bring down larger prey. Although, as we have learned, coyotes eat deer, antelope, and other large game rather rarely, circumstances of weather, terrain, or sickness among prey animals may encourage them to try their luck. Even a pair or a group of coyotes may have a hard time pulling down one of the larger hoofed animals, and a single predator can actually be in grave danger if it attempts to do so. But several coyotes, teamed against a sick or disabled deer or antelope, may have better luck if they combine their considerable skill.

The most interesting cooperative tactic consists of running the prey in relays. A pair of coyotes may hunt a jackrabbit this way, and can sometimes run even an antelope into the ground. Many prey species circle gradually when chased, and coyotes use this fact to their advantage. One of the dogs starts the animal while another lopes behind, conserving its energy. As the prey begins to veer to one side the back-up coyote cuts across the circle and takes up a position where it expects the prey to pass. When the chase reaches it, it springs to take up the pursuit, leaving its fellow to drop back and recuperate. Two or even three coyotes may employ this tactic, repeatedly cutting off the fleeing target until exhaustion puts it within their reach.

Although there is nothing unusual about wild dogs hunting cooperatively to pull down large prey, it is unusual for a dog to team up with an animal of another species. Yet there are well-documented instances, even photographs, of coyotes and badgers hunting together. The immediate question is why the coyote does not hunt the badger instead of hunting with it. It is probably because a single coyote would have a tough time killing an adult badger in good physical condition—and badgers appear not to be afraid of coyotes.

When coyotes and badgers have been observed hunting together the prey has been rodents or rabbits. The coyote may wait at the edge of a prickly pear clump containing a wood rat's nest while the badger digs out the rodent. If the rat runs out in the direction of the badger the coyote has gained nothing. But if the rat attempts to escape to the opposite side, as it often does, the coyote may snap it up. Or a badger may dig out and eat a rabbit run to earth by a coyote. The two animals do not go off on extended hunts together, and they do not share their kills. But the fact that the two species often poke about the places where small animals are found has led to a curious and mutually advantageous association.

Coyote social life is built around the mated pair. Animals tend to be monogamous during the mating season and some mated pairs stay together for several years. When an unmated female comes in heat she attracts a following of males, but once she accepts a male the relationship is stable and the bond social as well as sexual. Coyotes sometimes use a hole or other shelter to escape from heat or cold, but only during the reproductive season are they attached to a den. It may be dug from scratch, but often is the enlarged or cleaned out burrow of another animal such as a fox, badger, or another coyote. Coyotes often return to the same den from year to year if they are not driven from it by man, or forced to move to escape fleas or unsanitary conditions. The favored spot for a den is a well drained slope near the source of water essential for weanling pups.

The female coyote has a long heat period, a factor perhaps useful in building the social tie between the pair. The actual estrus, during which she is receptive to the male, lasts only three or four days. Estrus occurs in February and March in most females. During its height, sexual activity is similar to that of the domestic dog and wolf, with alternate scent-marking of objects, frequent mounting by the male, and copulation which includes the tie. The gestation period, like that of other genus members, is 60 to 65 days. Although the average litter has five or six pups, litters have been known to reach 19.

For the first three weeks after birth the pups nurse, and from the third week both parents also provide regurgitated food. Even before weaning, the parents bring young rabbits and rodents to the pups; after about six weeks, they no longer depend on regurgitation. The weanling pups emerge from the den as soon as their eyes are open, at around 14 days. They quickly begin puppy attempts to hunt, concentrating on insects as the easiest prey. Until they are old enough to accompany adults on the hunt, they keep both parents busy bringing food to the den, sometimes carrying it for miles. The successful rearing of a coyote litter seems to depend in good part on the male, which feeds the nursing female as well as the pups. If the female is killed after the pups are weaned the male often attempts to rear the young; if hunting is good he may succeed.

Young coyotes normally stay with the family during their first year although, rather than hunt as a whole family, they may break up into groups of two or three. The ccounts by early pioneers of large packs of coyotes are almost certainly exaggerated; when six or seven coyotes are seen together they are undoubtedly parents accompanied by offspring. (Larger groups come together around carrion, but they disperse after feeding and do not constitute true packs.) Coyotes become sexually mature at the age of two and until then it is not uncommon for some of the young to stay with their parents. Adults accept them and usually are aggressive towards them only during estrus. With the arrival of a litter unmated females sometimes act as foster-mothers, regurgitating food for the pups, playing with them, and even curling around them in the nursing posture. The yearlings which do not remain with the family gradually extend their roaming in search of unoccupied hunting territory and, during their second year, of a mate, though they may continue to come together with the parents for communal howls and occasional social contacts.

Coyotes living together establish hierarchical patterns, adults normally being dominant to young animals, and males to females. Conflict is similar to that between wolves, with the contesting coyotes assuming threatening postures and attempting to force down each other's heads or to knock each other off balance. Social bonds between females can be close enough to allow them to live together and raise their litters in the same den—even caring for each other's pups.

The coyote's most important natural enemy is probably the wolf. Because the two species are so closely related, wolves perhaps regard coyotes more as rivals than as prey and kill them when they get the chance. In the confined area of Isle Royale in Lake Superior, wolves may have accounted for the disappearance of coyotes. In more natural situations wolves probably do not have much effect on coyote members, though they may exercise a restricting influence and certainly kill occasionally. (Coyotes, of course, often profit by feeding on wolf-killed carrion.) Pumas and jaguars, like all large cats, seem to have a particular fondness for wild dog and kill and eat coyotes, especially trapped ones. Coyotes are so in fear of these cats that they avoid puma-killed carrion until the cat scent has dissipated.

A far greater check on coyote numbers is disease. They are susceptible to a variety of parasites, internal and external, and to the diseases which affect all members of the dog family. Although rabies is not common among them, it *can* occur in epidemics over large areas, as it did in the early part of the century in the western United States. Coyotes in California, Nevada, Oregon, Idaho, Washington, and Utah were affected and attacked livestock as well as some humans. Distemper also occurs among coyotes; pups are highly susceptible, though adults seem to build up a natural immunity.

Among internal parasites the most common—if not very harmful—is the tapeworm, the immature forms of which pass to coyotes which eat infected rabbits. Since rabbit is the staple diet of coyote pups, there is nothing remarkable in the fact that of 1,850 coyote remains examined in Kansas in the 1950's, 95 percent of the animals had tapeworms and appeared to have been infected as young as at four weeks old.

Other parasites are more serious. Round worms were found in the Kansas study to be present in 33 per cent of the animals; while not serious in adults, they can be among pups. The young animals are infected by the contaminated nipples of an infected mother, and if the infection is heavy the pups, like young domestic dogs, may die or become severely debilitated. Stomach worms in high concentration can also be serious to pups, though they are not important to adults. Hookworm is fairly common, the incidence rising sharply in areas where the animals live along river bottoms in warm climates. In dry areas like Utah it is almost nonexistent. Heavy infections in pups are often fatal.

Among external parasites of coyotes the flea is the most common; no coyote examined in the Kansas study was flealess. Curiously the coyote flea (*Pulex similans*) is a different genus and species from the domestic dog flea (*Ctenocephalus canis*). The only dog fleas found were on coyotes killed by dogs. Although fleas may disseminate diseases such as tularemia they do not appear to be particularly bothersome to the coyotes themselves, except in the case of animals debilitated from other causes or in connection with another condition such as eczema. Lice and ticks occur sporadically or seasonally but do not seem to bother their hosts very much either. Sarcoptic mange, though it is not widespread, can occur among coyotes; again, a heavy infestation would be serious for pups.

Adult coyotes are not only resistant to disease but also tough about injury. Numbers have been found which have survived and adapted to gunshot and trap wounds—they have had missing feet, broken legs, and deformed jaws. Such animals usually turn to livestock as easily available prey. As for skin wounds, a healthy adult appears well able to combat the risks of infection, for many animals have been examined which show healed cuts.

Man presents a different kind of challenge to coyotes. In the West, he is hated as a sheep-killer or simply a "varmint," reflecting the not-uncommon American attitude that any animal which cannot be eaten, milked, or ridden should be eliminated. Often the coyote is far more useful as a rabbit- and rodent-eater than he is detrimental to man for the sheep he kills. But the coyote has always been fair game and even today many states offer a bounty on the animal. A hundred years ago the largest coyote kills were by strychnine, which was put out for both wolves and coyotes and, while virtually wiping out the wolf through most of the U.S., reduced but did not eliminate the coyote. By the turn of the century the dangers of strychnine to man and livestock, as well as to all manner of wild animals, was at last being recognized and coyotes began to be hunted far more extensively by trapping.

Poisoning of coyotes, however, has not ceased. One method is by means of the cyanide gun or "coyote-getter." This device consists of an explosive cartridge charged with sodium cyanide, which is attached to a peg and hooded with wool or rabbit fur. The hood is smeared with a scent lure. When the hood is pulled, the cartridge explodes and shoots cyanide directly into the mouth of the animal, which is dead within five minutes. The cyanide gun, though somewhat more discriminating than strychnine, is a dangerous tool in coyote control, since the scent attracts not only other predators but occasional sheep and cattle. At that, it is not as pernicious as the ultimate poison, sodium fluoroacetate or compound 1080, which is added to small cubes of meat or fat and scattered indiscriminately for the delectation of all meat-eaters. The U.S. government is gradually awakening to the great danger of poisoning in predator control and in 1971 banned the use of poison by federal agencies in the United States. But private individuals and local governments remain free to poison coyotes and other animals as they see fit.

Like other members of the genus, coyotes are closely related to domestic dogs and will mate with them in captivity and occasionally in the wild. (The fertile offspring, called coy-dogs, can become a nuisance as stock-killers.) The fact that coyotes and dogs will mate leads to embarrassment for hunters proud of their coyote-killing hounds. Male hounds usually will not attack a female coyote in heat, and she seems to know it; for this reason hunters often include in their packs several female hounds which have no such scruples.

Like wolves, coyotes can become friendly with man if socialized young enough. Pups have frequently been raised in laboratories, or around ranch and farm homes, or have been fed while allowed to run free. Like other wild animals, individual coyotes, even litter-mates handled in the same way, vary in their degrees of tameness. Socialized coyotes show towards people whom they recognize the same friendly postures and movements that they show towards other coyotes—tail-wagging, lowering the hindquarters, sniffing, and mouthing. When more information is available on both wolves and coyotes, it should be very interesting to see if the somewhat different social patterns of these near-relatives result in any noticeable differences in the way the pups react to man and to their own kind.

THE GOLDEN JACKAL (*Canis aureus*)

Best known of the three species of Old World jackals, the golden jackal has a variety of common names, including gray, yellow, common, Asian, Indian and silver-backed jackal. The golden jackal resembles a wolf but is considerably smaller and lighter in weight. Its range is vast and lies for the most part to the south of the wolf's, across the Old World from north Africa to southeast Asia. It is found throughout North Africa from Morocco to Egypt, and south of the Sahara from Senegal in the west to northern Tanzania in the east. In southeastern Europe it occurs from the Adriatic through the Balkans and is encountered occasionally in Hungary and Rumania. The golden jackal is found in the southwest Soviet Union as far east as Armenia, ranges across the Near East, and is ubiquitous in India and Ceylon. It occurs in some sections of Burma and southwest Thailand, where it is not so plentiful as in India and may be a more recent immigrant.

The golden jackal varies physically throughout its range, the chief differences among the more than a dozen races being in size, color, length of coat, and minor tooth characteristics. Generally the fur is grayish-yellow, often with a reddish tinge. Along the back the tips of the hairs are black, but there is no black on the backs of the ears. The muzzle, ears, and legs are typically buff or reddish, becoming paler near the feet. Usually the underside of the animal, including the chin and throat, is whitish. The golden jackal may be as much as 16 inches tall at the shoulder; the length of the head and body is about 26 to 30 inches for males, one or two inches shorter for females. A large male may weigh as much as 25 lbs.; females rarely exceed 20 lbs. The tail is about a third as long as the body and head—proportionally shorter than that of the wolf.

The skull of the golden jackal resembles that of the wolf but is smaller and less bulky; the forehead is almost flat, with very little angle between the forehead and the muzzle. The jackal's teeth are comparatively small and the carnassials, especially, relatively weak.

As might be expected of an animal with so vast a range, the golden jackal makes itself at home in a variety of habitats. Not typically found in high or rugged mountains, it does live in hilly areas. Though not an animal of the most barren deserts it is found extensively in semi-desert areas. In the Serengeti region of Tanzania it prefers the open plains, or bush at the edge of the plains, to completely wooded areas. In Europe and the Soviet Union it seems most common in lowlands—among the reeds and bushes of river courses. It frequently lives near villages and populated areas and has made a name for itself as a scavenger, sometimes even of human corpses.

The golden jackal does not usually construct elaborate dens. Instead, it is likely to borrow those of foxes, badgers, or porcupines. If the jackal does dig one itself, it is content with a short, straight passageway dug into a thicket or under the roots of a tree. Sometimes it simply uses a depression in the ground. On the East African plains it often takes over old warthog holes which it further excavates; such dens usually have several entrances. It tends to be nocturnal, at least where it fears man.

The voice of the jackal, like that of the coyote, is one of its most frequently discussed features. The most common sound is a wail, repeated three or four times in an ascending scale, followed by three quick yelps, also repeated two or three times. This howl seems to be uttered most frequently at dusk or just before dawn. If one animal starts the howl it is taken up by neighboring animals and passed along. Like other members of the genus the jackal presumably howls as a social outlet, and to communicate its whereabouts. It also barks or yelps sharply, an alarm call uttered in the presence of man or a large predator.

The golden jackal eats anything it can get. It usually hunts alone or in pairs but will sometimes gather in a pack, probably made up of a pair and adolescent young, which is capable of killing fairly large game. Rodents and birds are the golden jackal's principal food, but it will kill reptiles, small deer and antelope, and small domestic animals. In some places, such as Kashmir, it is reported to kill kids and lambs. It is by no means averse to snapping up a straying chicken or other domestic fowl. The jackal is a great eater of insects, and specimens have been examined which had stomachs packed tightly with locusts.

The jackal will eat any kind of carrion or refuse and may be a regular visitor to slaughterhouses or garbage dumps. It also eats vegetable matter like berries, bulbs, watermelons and grapes. In the Tadzhik S. S. R. it feeds primarily on the fruit of the Russian olive in the autumn and winter, and in India eats maize, ber plums, sugar cane, and coffee beans. In Ceylon, golden jackals are seen trotting along the water's edge feeding on shellfish or edible debris tossed up by the waves.

A study of the feeding habits of the golden jackal in the Serengeti Plains of Tanzania yields some interesting information which, although not necessarily typical of the animal throughout its range, shows its adaptability as a predator. In Serengeti the golden jackal shares its territory with the black-backed jackal and tends to inhabit the open plain while its close relative is more common in the bush. The Serengeti is an unusual habitat because of the immense numbers of small Thomson's gazelle, known as Tommies, which also live on the plains. From January to April the gazelles are fawning and during these months the animals make up about half of the jackal's diet, with insects, especially the abundant dung beetle, constituting the other half. Eighty-one percent of the gazelles which the jackals were observed feeding on were animals they had killed themselves. Virtually all gazelles killed by jackals from January through March were less than two weeks old.

The diet before fawning time was quite different. From August to December gazelles made up only about one-fourth of the golden jackal diet, with insects comprising another 30 percent and small mammals, ground birds, carrion of wildebeest and zebra, and fruits constituting the rest. The difference in the number of gazelles killed by the golden jackal during these two periods is partly a function of the gazelles' fawning pattern, but it is also partly because of the jackals' own social pattern.

It seems that in the Serengeti, jackals rely on a running charge to capture their prey. They may occasionally stalk an animal but more often will start toward it at a trot which gradually increases to a fast gallop. The behavior of a threatened gazelle fawn is stereotyped. An older animal may run immediately, but a young fawn crouches low in the grass while the mother walks away a little distance, perhaps in an attempt to draw off the attacker. When the jackal is as close as ten yards or so, the fawn leaps to its feet and bounces away stiff-legged in the gait called "stotting."

Once the fawn begins to run the female gazelle attempts to defend it by trying to stay between it and the jackal, charging the jackal and butting at it, though usually not hitting it. Sometimes two or three female gazelles attack the jackal at one time. The

After Dorst and Dandelot (1970) for Africa; Hildebrand (1954) for Asia; and others

Golden jackal with fellow-scavengers. This jackal occurs in Europe, Africa and Asia. *Photo: George B. Schaller*

defense is often successful if there is only one attacking jackal, but far less so if there is a pair, for while the female gazelle attacks one the other can reach the fawn. Even when several gazelles are defending they all concentrate on one predator, leaving the other open to make the kill. The increased success of a pair of jackals accounts in part for the rise in gazelle kills during the January-to-April period. It corresponds to their own breeding period and a combination of tandem hunting plus the availability of young fawns obviously means more kills to feed jackal pups.

The mating season of the golden jackal is January and February, the gestation period from 60 to 63 days. The usual litter is four or five pups, though as many as nine can be born. (On the Serengeti Plain, however, a smaller litter of from one to three pups seems to be the most common.) There is a strong pair-bond between the parent jackals; the male feeds the pups as well as the female while she is confined to the den. For the first two weeks jackal pups remain in the den. After their eyes open and they can crawl or walk about, they begin to emerge to play about the entrance and to make their first attempts to catch insects. From about two weeks on, they are fed regurgitated food brought by the adults. This they solicit by running to the approaching adult to dance about it excitedly, keeping their muzzles near the corner of the adult's mouth and wagging their tails vigorously while the parent regurgitates chunks of meat. If the female has remained at the den, she too solicits food from the male, and the pups sometimes then beg from her.

Young jackals grow rapidly; by the time they are four or five months old they are almost as big as their parents. They are able to catch termites and dung beetles, but are still dependent on their parents for the main part of their diet. From about three months on they may accompany the adults on the hunt but their efforts are initially more hindrance than help and only gradually do they become accomplished hunters. Jackals usually reach sexual maturity at the beginning of their second year.

The golden jackal's close contacts with man have led to its figuring in countless stories and myths, and to its being both persecuted and worshipped. The jackal-headed god of the Egyptians, Anubis, may have been given that head to placate the jackals which ravaged the necropolises in the desert. The jackal may have been worshipped in ancient Egypt, but in the Soviet Union today it is regarded as vermin, on the grounds that its scavenging of disease-ridden carcasses spreads disease and that it is a severe scourge both to domestic fowl and to wildlife. It is also hunted there and in Europe for its pelt, though the hair is short and coarse. In colonial India the British hunted the jackal with hounds, since it was swift enough to give a good run.

In Ceylon the golden jackal is regarded as the wily one, an animal which outwits others. In India it is linked with the tiger, and in the Middle East with the lion, as a friend and sentinel of the larger predator. Such stories may have some basis in truth, for certainly the jackal follows both tigers and lions to feed at their kills, and jackals have been known to give an alarm call when man approaches, thus earning their keep.

THE BLACK-BACKED JACKAL (*Canis mesomelas*)

Of the two strictly African jackals, the black-backed one is the more common. It is a conspicuous scavenger in the great game areas of East and South Africa, and in South Africa, where its natural prey has been eliminated, it has learned to kill sheep. In various parts of its range the black-backed jackal is called silver-backed, silver, saddleback, silvertail, red, and even golden jackal. It ranges from Somalia, Ethiopia, and southern Sudan south to the tip of Africa. It is found as far west as the Uganda-Congo border, South West Africa, the Kalahari Desert, and southwestern Angola. In East Africa, where it shares its range with the other two jackals, it is the most common of the three species.

The black-backed jackal's most distinctive feature is its marking and color. It has a dark saddle-shaped marking on the neck and back, broadest at the shoulders and becoming narrower over the back until it reaches a point at the base of the tail. The hairs of this saddle are mixed black and silvery white, darker in some animals, lighter in others. In distinct contrast to the back, the sides of the head and the outer sides of the legs are reddish. The underparts of the jackal—throat, chest, belly, and groin—

range from sandy to almost white. The tail, which is moderately bushy, is brownish with black-tipped hairs on the upper side and a dark or black tip. It has distinctively long, triangular, pointed ears, and its narrow head and pointed muzzle give it a somewhat foxy appearance.

Longer and taller than the golden, the black-backed jackal varies considerably in size in different parts of its range. The length of head and body have been described as anywhere from 36 to 43 inches in East and South Africa, to only 28 inches in Somalia. The tail is 12 to 14 inches long. The animal may stand as high as 19 inches at the shoulder and usually weighs 20 or 21 lbs. When trotting it carries its head a little lower than the shoulders and its tail extends out and downward with the tip drooping. However, when feeding or excited it may raise its tail above the line of the back, or even wave it high in the air

The black-backed jackal has several different calls or howls. When hunting, or in the evening before a hunt begins, it howls or yelps a drawn-out sound followed by several short ones. Rendered by the natives of East Africa as "Bweha! bwe-bwe-bwe-bwe!," this call has led to the animal's Swahili name of *bweha*. The call may be communication to a mate or family members; in some cases it is apparently excitement over a kill made by another predator. The jackal also has an alarm call which sounds something like "ke-ke-ke-ke-kek," given when it is startled or cornered, or when large carnivores are on the move. The female utters a low growl to call pups from the den, and either parent may produce a soft "wuf" if it becomes aware of danger near the pups.

An animal of the open plains and scrubland, this jackal is not found in dense jungle, forest, or mountainous areas. Although it uses some sort of den or burrow at whelping time, during the day it prefers to lie in tall grass or under bushes rather than take shelter in a burrow. For much of the year animals associate in pairs, and sometimes hunt in family groups. Many unrelated animals will come together to feed on a kill or carrion, but then disperse.

It is a rather strongly territorial animal and once a pair chooses an area they are inclined to stay there. Other jackals respect this territory and are unlikely to live too close, so that jackal pairs will be distributed fairly evenly according to the amount of food and shelter available. On the Serengeti Plain of Tanzania the average territory is about two miles in diameter, a figure which might vary widely elsewhere.

In many places where the jackal now lives it is nocturnal, leaving its burrow or resting site at dusk and returning at dawn. This pattern may very well be the result of pressure from man, for where the jackal feels relatively safe, as in game parks, or when a nearby kill lures it, it will instead be up early and return later. In the game preserve of Kruger Park it reportedly does not distinguish between night and day.

The black-backed jackal will eat almost anything. It both scavenges and hunts for itself—as a scavenger following lions, hyenas, and other larger predators, and as a hunter running down small or young antelope, rodents, birds, reptiles, and insects. The percentage of its diet filled by various foods varies widely depending on area and season. As a scavenger the jackal must take a position behind the scavening lion or hyena, competing with vultures and marabou storks for third place at the table. Feeding lions are commonly ringed by black-backed jackals waiting at a respectful distance.

As a hunter, the jackal is a good stalker, and puts considerable time and patience into watching a sleeping or unwary ground bird—snatching the prey off the ground or leaping into the air if it is flushed prematurely. In addition to catching birds on the ground it eats eggs, including those of the ostrich, a profitable undertaking considering the size of an ostrich egg. A jackal may also spend hours in front of a springhare burrow, waiting for the occupant to emerge. It will scuffle through the grass in search of mice, or dig them out; sometimes two or more jackals cooperate in mouse-hunting, one digging while the mate or young stand by to snatch up what emerges. The black-backed jackal eats insects as well as fruits and berries (in farming areas of East Africa it is reported to be especially fond of strawberries).

The jackal also hunts larger game, but is restricted by its size to the smaller antelope or the new-born young of the larger ones. A study of the feces of the black-backed jackal in the Serengeti Plain shows that from August to December, when the jackals are

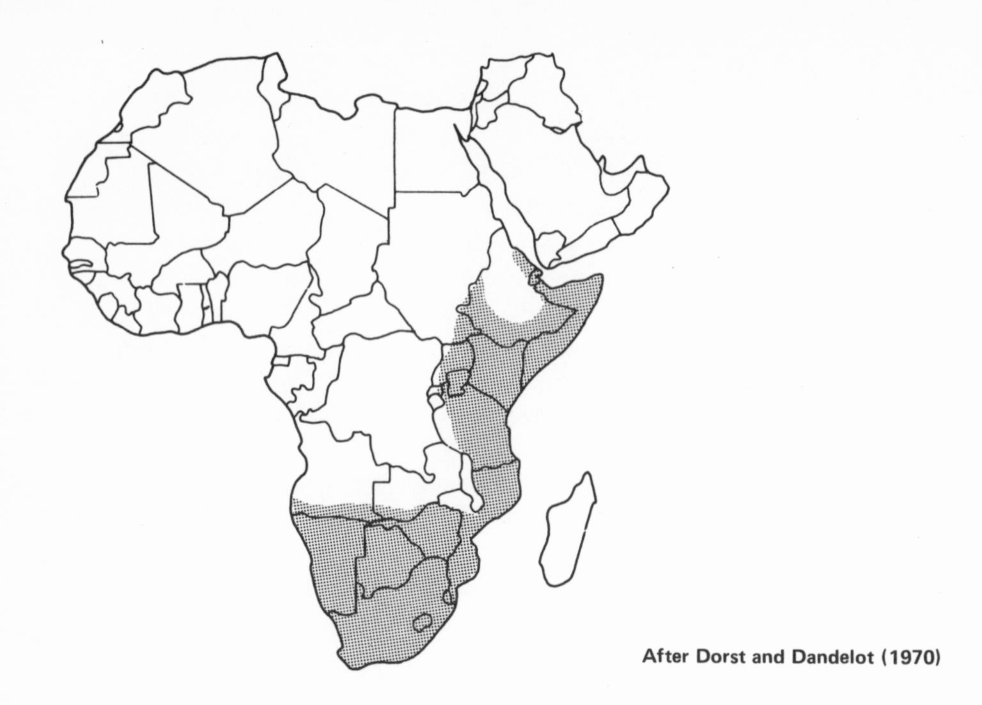

After Dorst and Dandelot (1970)

Black-backed jackals of Africa feed on a zebra foal dead of disease. *Photo: George B. Schaller*

breeding and raising their young, about half of their diet is composed of the small Thomson's gazelles, most of which they evidently kill themselves. The remainder of the diet at this time of the year is made up of small mammals, ground birds, carrion, and vegetable matter. From January to April, fawning gazelles continue to provide about half of the jackal diet but the rest is made up mostly of dung beetles, which are abundant in the droppings left by large concentrations of grazing animals.

The black-backed jackal most successfully kills small gazelles when it hunts in pairs or family groups. A pair of jackals can also prey upon the very young calves of larger antelopes, such as the wildebeest, by cooperating to lure the female away so that the calf is left unprotected. When the jackal kills sheep it uses the same method as in hunting antelope, running in among the group, chasing a victim and pulling it down. Here too a pair of jackals or a family often hunts together.

Not surprisingly, a comparison of jackal predation in the Serengeti, where Thomson's gazelles are numerous, with the nearby Ngorongoro Crater, where gazelles are far outnumbered by the larger zebra and wildebeest, indicates that in Ngorongoro carrion is far more important to the black-backed jackal diet. In the Crater, carrion composed about 35 percent of the contents of jackal feces, compared with some 3 percent in the Serengeti. In Ngorongoro the black-backed seems to be a far more successful carrion-feeder than its cousin the golden jackal. Although golden jackals in the Crater outnumber the black-backed almost three to one, the latter are as common around hyena kills as the former and are more likely to move in among the hyenas to feed than to hover at the edges waiting for the hyenas to retire, as the golden jackals do.

A fascinating account by Randall Eaton of cooperative hunting between black-backed jackals and cheetahs in Nairobi Park indicates the jackal's adaptability. Eaton's observations involved a pair of jackals with three pups, and a cheetah with four cubs. Seven times during the weeks in which he studied cheetahs in the Park this one family of jackals facilitated the stalk of the adult cheetah, which killed successfully in two of those hunts, giving the cheetah a success ratio twice as good as that observed when it hunted alone. The jackals would run to a herd of antelope, such as impala or Grant's gazelles, which were near to where the cheetahs were lying in wait. By their barking and running back and forth the jackals distracted the game. As soon as the cheetahs saw that the antelope were preoccupied they moved towards the game and were able to get quite close before being spotted. The cooperation, which was restricted to these two families, had probably first occurred at some time when the jackals were moving through a herd barking, as they often do, and the cheetah, hunting near by, had been able to make a kill as a result.

When hunting, the black-backed jackal moves at a trot, a pace of five or six miles an hour, and prefers to follow paths, roads, or natural features of the terrain which allow easy traveling. It can keep up a steady pace for hours and may cover 25 miles in a single night if the pups are demanding and food is scarce or far away.

Jackals tend to be monogamous, sometimes from year to year. They choose their mates a month or two before the female actually comes in heat and, like other wild dogs, young jackals frequently choose mates from their own litters. In South Africa, mating takes place from May through July. Since the gestation period is 60 to 65 days, whelping occurs from July through September. In Tanzania most mating takes place from mid-July to late September, and in South West Africa occurs a little later. A maiden female produces her first litter late in the season, since she chooses a mate at nine or ten months of age, comes in heat at perhaps 12 months, and whelps two months later.

A female jackal seeking a mate attracts a following of several males. She urinates frequently and the following males urinate in the same place and kick dust. The males often fight fiercely among themselves and the female will usually choose the dominant one. The actual mating results in the tie typical of dogs.

After mating the female jackal begins her search for an appropriate den. Jackals use burrows but do not usually dig them themselves, preferring to clean out and enlarge an old aardvark hole or that of some other animal, or to adapt a convenient cave or crevice. In the Serengeti, a dug-out termite mound seems to be the most popular den site. The litter size depends on several factors, one being the age of the female. Though most sources list an average litter size of three to six or seven, old females and those whelping for the first time usually have one to three pups, while females in their prime may have from four to as many as ten.

When the pups are born the female remains in the den constantly for the first three days, and for the next week stays with the pups except when out hunting at night. While she is den-bound, the male brings her food. After about the tenth day, when the pups' eyes are beginning to open, she leaves them alone most of the time and only enters the burrow to suckle them. For the first weeks the pups are dependent on milk, but from about three weeks on the parents also feed them regurgitated food and bring them scraps of meat to chew on and play with.

Black-backed jackal pups begin to emerge from the burrow at about two weeks of age, when their eyes are fairly well open and they can crawl or stagger about. Like other young dogs they are very playful, dashing and tumbling about the den entrances so that little footpaths are worn to neighboring bushes or rocks where they play hide-and-seek. At this stage the pups are a juvenile lead-gray all over, but by three months they have the black and white saddle and reddish limbs of the mature animal. They grow rapidly and by the time they are completely weaned, at eight or ten weeks of age, are ready to accompany the adults on hunting trips; within two or three more weeks they will leave the den to live on the open plains with the parents.

When young jackals first begin to accompany their parents they are not much good as hunters and get in the way more than they help, but by the age of six months they are nearly full-grown. The family stays together until the pups are eight or ten months old. At this age young jackals are pairing off within the litter. They begin getting the urge to wander and usually leave the mother of their own accord; occasionally, however, yearling jackals stay with the parents as part of an extended family. When the pups are six or seven months old, the father often leaves the female for a period of wandering on his own, but apparently in some cases the two animals find each other at the next mating season.

The black-backed jackal is susceptible to a variety of diseases and parasites, some of which are significant to man and his domestic animals. It can suffer from rabies, distemper, biliary fever, nagana (the sleeping sickness fever caused by the tsetse fly), anemia, Rickettsiosis and toxoplasmosis. It has three kinds of fleas, at least two of which are vectors of plague. In South Africa it carries four kinds of ticks, one of which is a carrier of biliary fever and Rickettsiosis and another of which causes paralysis in sheep. The jackal also has the usual internal parasites common to the dog family.

In the settled parts of Africa, the black-backed jackal has incurred the wrath of man, not because of the diseases it may sometimes transmit, but because it may become a sheep-killer. As usual, man himself has caused the problem. As the lion disappeared from South Africa, so did the lion kills on which the jackal fed; as the antelope were shot off, the jackal lost another of its normal sources of food. Poisoning and gassing of rodents, and spraying against insects, reduced the numbers of smaller fry on which the jackal traditionally depended, and the encroachment of man on the natural habitat likewise reduced the availability of small mammals. Jackals turned by necessity to killing sheep; no longer subject to pressure from their natural enemies, which had been killed off, they increased to plague proportions.

In many cases the control measures adopted against the jackal have made the problem even more severe: poison inadvisedly put out for the jackal, for instance, which also kills animals on which the jackal feeds. One reason for the plague of jackals in South Africa may be that inbreeding, which limits fertility and viability, has been discouraged by attempts at control. Under natural conditions, as mentioned, jackals often find mates in their own litters, but if persecuted, they tend to scatter, thus mixing gene pools. The question is why under natural conditions a system of inbreeding would have developed at all. The answer might be that it is a mechanism for keeping the jackal population stable and not allowing it to exceed its food supply. Perhaps pairing between litter mates, which may enhance the social bond between adults, helps offset the genetic disadvantages of inbreeding. However that may be, with the interference of man, although the unwary and unfit among the jackal population are killed, the survivors are tough, intelligent, and above all prolific.

In addition to man, the black-backed jackal has a number of natural enemies. Members of the dog family, and jackals are no exception, seem to be uniformly attractive to leopards. In some places, the python frequently kills jackals, apparently because the jackal's habit of traveling on footpaths brings it within reach of the waiting snake. The large birds of prey frequently attack jackal pups asleep or playing outside the den. Baboons also kill jackals, since the jackal sometimes attacks their young and if caught by an angry adult is not likely to survive the encounter.

THE SIDE-STRIPED JACKAL (*Canis adustus*)

Judging by the amount of attention paid it, the side-striped jackal is the poor cousin of the jackal group. Far less often seen than the golden and black-backed jackals, it comes in close contact with man less frequently. Also called the gray or gray Rhodesian jackal and in Afrikaans *witkwasjakkals,* which refers to its white tail tip, the side-striped jackal is an animal of the open woodlands and savannas of central Africa. It ranges from Senegal east to Sudan, Ethiopia, and Somalia in the north; south to northern South West Africa, north Botswana, and the north parts of Zululand and the Transvaal in the Republic of South Africa. It may gradually be extending the southern edge of its range.

The side-striped jackal is larger and looks somewhat more wolf-like than the other jackals. It stands 20 inches at the shoulder, with a body length of 25 to 31 inches, and a tail from ten to 14 inches long. Males weigh as much as 30 lbs. and females average about 24. The coat is a grizzled or brownish gray, somewhat darker on the back and blending to tan on the chest and limbs, but with none of the distinctive saddle of the black-backed jackal. The tail tends to be darker than the body, is rather bushy, and usually ends in a distinctive white tip.

From shoulder to hindquarters on each side runs a white stripe, and right below it a black stripe. It is these stripes, very distinct in some animals yet almost nonexistent in others, that

The side-striped jackal, found only in Africa, looks more wolf-like than other jackals. *Photo: Pretoria Zoo*

have given this jackal its most common name. The muzzle and face are broader and less foxy than the black-backed jackal's; the ears are relatively short (three inches or less) and slightly rounded at the tip. The backs and edges of the ears are the same grayish-brown as the back, but the long white hairs on the inner surface make the ears look white from the front.

The side-striped jackal prefers thick woods or bush to the exposure of the open plain, and favors a habitat near water. It seems to be much more nocturnal than the other two jackal species, but will come out in the daytime if the weather is cool and cloudy. Shy and somewhat sluggish in disposition, the side-striped jackal is much less noisy than its close relatives. Its voice is a series of single barks, lower in pitch, more "metallic," and uttered more slowly than the black-back's rapid cry.

The diet of the side-striped jackal is made up of carrion to an important degree. But though it is thought of largely as a scavenger and is sometimes seen at lion kills, it also eats rodents, birds, insects, reptiles, berries and other fruit. Although it eats fewer antelope than the black-backed, the side-striped jackal is perfectly capable of bringing down the smaller grazing animals. In South Africa it does not kill sheep, perhaps because its timidity and preference for a secluded habitat have simply kept it away from temptation.

Its gestation period is 60 to 65 days. Where there are seasonal changes it whelps in the wintertime, which in southern Africa means from June on. Litter size averages from three to seven. It bears its young in burrows and in typical jackal fashion prefers to use those dug initially by other animals, especially the useful aardvark. Though the side-striped jackal is reported to interbreed with domestic dogs, we know nothing about the offspring of such matings.

THE DINGO (*Canis dingo*)

The Australian dingo or warrigal looks and acts like a typical wild dog, preying upon small grass-eaters and harassed by man for its sheep-killing propensities. Its peculiarity lies in its historical relationship to the fauna of Australia: the dingo is the only carnivore on the continent (excepting those introduced by European man) and, unlike most indigenous Australian mammals, it is not a marsupial. This anomaly is explained by the fact that it is actually a feral animal, brought to Australia by primitive man as a domesticated or semi-domesticated dog. It must have reached Australia quite early, since its fossil remains have been found with those of late Pleistocene marsupials. Dingos found in Australia an advantageous environment and spread rapidly over the continent.

Some of the dingo's effects on prehistoric Australian fauna can only be guessed. But most authorities feel that by its efficiently pre-empting the same ecological niche as the Tasmanian wolf (a large marsupial predator), it was probably instrumental in causing the disappearance of that animal from the Australian mainland. Whether the dingo also seriously affected the marsupial grass-eaters which it encountered upon its arrival is more difficult to say; however, it is certain that when European man further upset the natural balance by his importation of sheep and his introduction of the rabbit, the number of dingos increased rapidly in response.

To save his sheep, man then began exterminating the dingo, a program which benefited no animal so much as the rabbit. The dingo, however, survived, and when myxomatosis was introduced in the 1950's to eliminate the rabbit, the dingo turned in earnest to killing sheep. The dingo-rabbit-sheep story in Australia is a classic and glaring example of the unhappy results of a long history of man's introduction of exotic species into a balanced ecology.

Despite its primitively domestic origins, the dingo is now considered a wild dog. The size of the average animal is between that of a wolf and a jackal, its height varying from 19 to 23 or more inches at the shoulder. From nose to tip of tail, it is about five feet long. Adults average from 50 to 70 lbs., although individuals as heavy as 120 lbs. have been recorded. The dingo's ears are relatively short, somewhat rounded at the tip and broad at the base, and carried erect. The tail is well furred, even bushy, and is carried low when the animal is at rest, raised when it is

This barking Australian dingo looks very much like its close relative the domestic dog. *W. Nicol, Australian News & Information Bureau*

active. The thickness, length, and consistency of the dingo's coat depend on the climate. The animals which inhabit tropical northern Australia have short flat coats with little or no undercoat. Those from mountainous or southern areas may have a thick coat with coarse outer hair and soft under-fur.

The dingo is usually tawny, darker on the back and lighter to buffy-white on the undersides and throat. The feet and tip of the tail are often white. The tawny tone may vary from very pale to quite reddish. Dingos are not necessarily tawny, however, and variations in color are not proof of hybridization with domestic dogs. Early accounts by European settlers and explorers in Australia make it plain that the pure dingo may range from white or even albino through all phases of tawny or brown to black. Piebald and black and tan dingos were reported by early writers, and a blackish-brown brindling is not uncommon.

An interesting comment on dingo color, and incidentally on the affection of the Australian aborigine for his tame dingo, is found in the diary of a Major Lockyer who visited Stradbroke Island, off the coast of Queensland, in 1828. "The attachment of these people to their dogs is worthy of notice. I was very anxious to get one of the wild native breed of black color, a very handsome puppy, which one of the men had in his arms. I offered him a small axe for it; his companions urged him to take it, and he was about to do so, when he looked at the dog and the animal licked his face, which settled the business. He shook his head and determined to keep him."

When Europeans arrived in Australia the dingo was widespread over the continent—in habitats that varied from plains to forests and mountains. The vigorous campaign waged against it by man has pushed it out of certain areas, especially the eastern, more populated parts of the country. Dingo strongholds are now the forests of the Great Dividing Range, northern and central Australia with their small, scattered human populations, and the arid regions of South and West Australia.

The dingo's voice, generally a yelp or howl, is a means of social interaction; a family group sometimes howls together before starting out on a hunt, and during the pursuit, may howl to keep in contact. The howl itself begins low, rises to a very high note, and then descends. Like other dogs, the dingo has many different howls, yelps, and barks used in a variety of circumstances.

Dingos, like jackals and coyotes, hunt singly, in pairs, and in family groups, but only rarely in packs, though adults or pairs may share a home range. Where unmolested, they are diurnal, though they rest in the heat of the day, but in areas close to man they prefer to be abroad at night or at dawn and dusk. Like most dogs, dingos eat a little of everything, including reptiles, insects, rodents, ground birds, carrion, and vegetable matter. Though originally the staples of the dingo diet were kangaroos and wallabies, the disappearance of many indigenous animals has caused dingos to depend heavily on rabbits and sheep.

They cannot run particularly fast, top dingo speed being about 30 m.p.h., so they depend on endurance, encirclement, and a variety of ruses to catch their prey. Techniques practiced by other dogs for killing large prey are used by the dingo in hunting emus, large kangaroos, and cattle, all of which can kick with deadly effect. Two or more dingos attack simultaneously, one distracting the prey from the front while another attacks from the rear. After damaging the hindquarters of a cow or kangaroo, they may wait for the animal to stiffen and weaken before completing the kill. Dingos have also been seen using a variety of antics to try luring a young calf away from its mother.

Where their natural prey has disappeared altogether, dingos are very destructive to sheep, and their presence in any numbers makes open range sheep grazing impossible unless countermeasures are taken. A family of dingos may race among a flock at night, slashing and biting until a dozen or more animals are killed and as many wounded. It is often suggested that sheep somehow excite a "blood-lust" in the dingo which causes this wanton slaughter. More likely it is a combination of the utter defenselessness of sheep with the excitement and lack of expertise on the part of young dingos that results in far more damage than would occur with wild prey.

What little is known about the social life of the dingo in the wild comes mostly from the people who hunt it—sheep raisers and trappers. It has bred regularly in the London Zoo for many years, however, and biological details of reproduction are well known. It usually mates once a year and the litter is born in late winter or early spring, from August through November in Australia. (In the London Zoo the majority of animals conform to the seasons of the northern hemisphere and produce their young between February and April.) The gestation period averages 63 days and litter size ranges from one to eight or more with an average of four or five. Young dingos generally mate first at the end of their second year. Evidence that mate preferences may remain constant over several years was given by a female dingo at the London Zoo which showed a decided preference for her mate of the previous year, even though he was separated from her by a fence and two other males in her enclosure mounted her persistently.

To raise its family the dingo prefers to borrow an abandoned rabbit warren rather than dig its own, though it may also elect a cave or a hollow log. It may choose several possible dens, and after the litter is whelped, moves the young from place to place if danger threatens. Usually both parents cooperate in caring for the pups, which begin to eat solid food at about two months old and shortly thereafter join the adults in hunting. The young normally stay with the parents for about a year, occasionally for two or three.

As might be expected of an extraordinarily successful exotic species, the dingo has few enemies other than man. Some dingos fall prey to the carpet snake, the python, and the crocodile in the tropical areas of Australia, and the wedge-tailed eagle, a very large bird, takes a toll of young dingos. Man is the great enemy, or rather the white man, for the dingo occupies a favored status with the aborigine, who captures the pups, raises them to be devoted dogs, and trains them to hunt. The aborigines are fond of their animals—to the extent that women sometimes suckle orphan pups. To a primitive hunter like the aborigine, whose existence is marginal and precarious, a trained hunting dog is an invaluable asset, well worth such care and attention.

The white sheep-raiser, on the other hand, has been waging a relentless war against the wild dog ever since he arrived on the continent, using traps, rifles, hounds, and poison; the attack was so intense in the 19th century that it was assumed the dingo would soon become extinct. Nothing of the sort happened and the harassment continues. All of the Australian state govern-

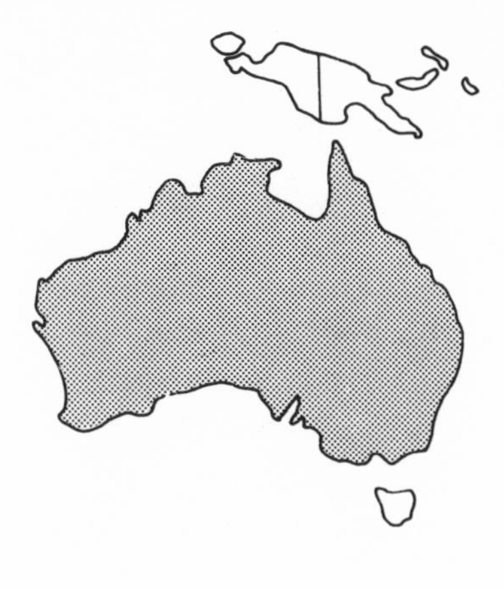

Female dingo, tail raised in excitement, supervises her pup. *Photo: Australian News & Information Bureau*

ments employ doggers and offer bounties although, as elsewhere, the bounty system is fraud-ridden and less effective than it is supposed to be. A price of at least four Australian dollars (U.S. $4.50) is paid by every state for a dingo scalp, with the bounty increased to 50 times that for notorious individual animals. The number of dingos taken is enormous; in Queensland alone the government paid bounty on about 30,000 dingos a year during the 1960's.

Another way to fight the dingo is to try to fence him out. Queensland by 1961 had completed a 3000-mile section of an anti-dingo fence along its northern and western frontier which is designed to keep the dingo beyond grazing lands. (The total length of all sections of dingo fence in eastern Australia is about 6000 miles.) The fence is six feet high and is sunk a foot into the ground, but since dingos wiggle through gaps in the fence, burrow under it, and climb trees beside it to jump over, all vegetation and sand must be bulldozed away in a track on either side, and the fence must be patroled regularly. The cost of the Queensland fence was over $25 million but it encloses some 80 percent of the state's sheep country and should cut dingo losses considerably.

Since 1946 a new and ecologically horrifying method of control has been the use of poisoned bait dropped from the air. Generally strychnine has been used, a dose of one-half pellet to a fat cube, and approximately one million a year were dropped along the Queensland barrier fence and in remote areas of the state. In 1968 the new poison called compound 1080 (sodium fluoracetate) was tried in Queensland. From May to August of that year, 750 landholders participated in the project and provided 150,000 pounds of fresh meat to be treated. From this campaign the Queensland government officially concluded that 1080 poison is one of the most effective means ever used in Queensland against dingos. Judging from the results of this compound in the United States and elsewhere it must be assumed that the Queensland campaign was highly effective against untold number of other creatures as well.

The dingo is biologically very similar to the domestic dog. The skull and teeth strongly resemble the oldest fossils of domestic dogs of Europe, but are more wolf-like than those of modern dogs, especially in the size of the teeth. Not surprisingly, the dingo mates readily with the domestic dog to produce fertile hybrids. Crosses between dingo and dog have sometimes resulted in especially destructive and cunning sheep-killers; the so-called Red River dog, for instance, accounted for over 1000 sheep and nearly as many cattle in the ten years before it was destroyed. The dingo is also supposed to have contributed its genes to at least two of the native Australian breeds of domestic dogs, the blue cattledog known as the Heeler, and the Australian sheep-dog or Kelpie. Both of these breeds appear to be the result of Dingo-Smooth Collie crosses which were then backcrossed to the Collie.

Despite the constant preoccupation with the dingo on the part of Australian stock-raisers and governments over the past 100 years, there has never been a comprehensive study of dingo biology. Virtually all information published about the animal to date has been concerned with its control. This lack should be met by a study currently going on under the sponsorship of the Division of Wildlife Research of the Commonwealth Scientific and Industrial Research Organization (CSIRO). It will be some years before the results of the study begin to appear, since field work on the life of an animal like the dingo is a slow and laborious process. But those interested in wild dogs—or, more broadly, in the ecological balance of a unique continent—will await eagerly what should be the first complete, concrete, and unbiased report on Australia's arch-predator.

THE DOMESTIC DOG (*Canis familiaris*)

This chapter on the domestic dog cannot follow the same format as those on other species. It would be pointless to describe size and color of what is unquestionably the most varied animal species in existence. To say that the domestic dog can range in height from six to 40 inches at the shoulder and can weigh from four to 160 lbs., or that it may occur in almost color or coat pattern known among mammals, hardly pinpoints its characteristics.

We will deal here, instead, with the origin of the dog and what is known of its domestication, will mention several early types, and will note the existence of the curious feral, or perhaps more accurately semi-domesticated, dogs called pariahs. We will also discuss how the domestic dog differs from its wild ancestors and how domestication has affected various physical characteristics and types of social behavior.

We know that the domestic dog is more closely related to the wolves, coyotes, dingos, and jackals—that is, to other members of the genus *Canis*—than it is to other genera of the family. But from which of the wild members of the genus is the domestic dog descended, or is it the descendant of a species now extinct? In answering the question of origin we have two sources of information—historical and archeological information on the process of domestication, and knowledge of the anatomy, physiology, and behavior of the species concerned.

Most widespread of all domestic animals, the dog has been known from the dawn of history in Europe, the Mediterranean, the Near East, and Asia. Domestic dogs were found throughout tribal Africa in forms which existed before the advent of colonialism, and also occurred from one end of pre-Colombian America to the other. A part of all major civilizations in recorded times, dogs have obviously been recognized as desirable additions to society; people all over the world have been eager to acquire domestic dogs and transport them during mass movements and migrations.

This is proof not only of the dog's ease of transport, but of its universal adaptability. It can withstand the severest arctic conditions and pull far more than its own weight at the same time. It can travel conveniently in small boats and canoes over long distances and serve as an emergency food supply to boot. It can go under its own power over virtually any terrain where man can pass, acting as a hunting companion, guard dog, and pack animal. The dog can survive on the most pedestrian or meager diet, making do on nothing more than the left-overs of its master.

No one knows when or where the dog was first domesticated. The archeological record is not complete enough to supply answers to the myriad questions involved. Some of the oldest dog remains come from Denmark, in a culture level transitional between Mesolithic and Neolithic called the Maglemosian. These remains may be as old as 10,000 B.C. or may date from as recently as 6000 B.C. From the Near East the earliest authentic dog remains are from Jericho and can be dated approximately 6500 B.C. Nothing more ancient than these finds can be considered unquestionable.

More recent evidence on canine domestication is easier to come by, once again primarily from Europe and the Near East. We have an Egyptian bowl dating from about 3500 B.C. with a picture of four dogs of the greyhound or saluki type. Skeletons of the same date have been found in Mesopotamia; the bones have the elongated body and limbs of the saluki. The existence of such highly refined types indicates that dogs had been bred systematically for some time, probably first in Mesopotamia and spreading from there to Egypt. Archeological and historical writings from that date on yield frequent references to dogs. They were used in the Near East for a variety of purposes: hunting, guarding, and as pets. Various types of house dogs retaining such juvenile or mutant characteristics as a tightly curled tail, short legs, and shortened muzzle were developed. In some cases dogs were mummified and buried in special tombs or with their masters.

In pre-Roman Europe, the record concerning dogs is based primarily on bone remains associated with a variety of Stone and Bronze Age sites. New Stone Age (c. 4000 B.C.) dog remains have been found near Swiss lake dwellings and in the Novgorod region of Russia, as well as in Scandinavia, Germany, France, and northern Italy. From the coming of the Bronze Age (about 3000 B.C.), dog remains are more and more numerous and occur from Ireland to Spain and east throughout Europe, indicating that the dog had by this time become an integral part of prehistoric society. With the rise of Greek and Roman civilizations, dogs begin to appear on coins, statuary, bas-reliefs, and finally in classical writings.

Dogs in the New World appeared more recently than in the Old. The earliest evidence for North America is placed as far back as 5000 B.C. There seems to be no question that the dog was brought from Asia by the peoples who populated the Americas. There is no indication of a close relationship between the domestic dogs of prehistoric Indians in the Americas and the native American wild dogs.

Although we know that as early as five or six thousand years ago dogs of several different distinctive types had been developed in the Near East and Egypt, there was evidently less, if indeed any, systematic breeding of dogs in Europe at the same time. Earlier archeologists attempted to sort out by type, and to assign names to, the various dog remains from Neolithic sites. More recent workers, however, have found that European Stone Age dogs were all very similar compared with the great variations in modern dogs, and can be described generally as medium-sized, dingo-like animals. They were relatively uniform in size and their teeth had different proportions from those of local gray wolves (from which they were not immediately descended).

As the Bronze and Iron Ages succeeded the Stone Age, European dogs gradually increased in size. The increase was probably in part an expression of the closer ties between dogs and people, which resulted in improved diet for the dogs. Also, it was almost assuredly the result of a certain amount of selection by those peoples, especially northern, who used the dog as a draft animal. This selection might have been unconscious, the result of favoring larger and sturdier animals, feeding them better, and caring for them more closely.

In places where successive waves of immigrants brought with them successively larger animals, these bigger animals did not represent completely new types of dogs, and were certainly not the result of the domestication of different wild dogs. Work on prehistoric dogs in the Americas, for instance, shows that the smaller types belong to older cultural levels, with more recent groups having progressively larger animals. In operation here was a principle common to many widely dispersed species: individuals living closest to the center of origin of the species tend to be more modified than those at the fringes of distribution, which tend to retain the original type. Thus peoples moving northward in Europe and Asia were bringing with them larger types of dogs from the original centers of domestication.

The increasing size of European dogs was a conspicuous trend in Stone and Bronze Age cultures. As peoples became more sophisticated, first in the Near East and then in Europe, another and sometimes simultaneous trend was occurring—increasing diversification of dog types. This seems to have been the result of conscious interference by man. It is marked in the development of the slender, long-legged, presumably sight-hunting variety of the greyhound type which so early appeared in Mesopotamia and Egypt. Diversification was also obvious in the appearance, in the Near East, Europe and Asia, of small dogs which today we would call lap dogs. And it was also marked by the retention of mutations we have already mentioned: the curled tail, extreme shortleggedness found in breeds like the dachshund and basset hound, the bulldog type of short face, and the lop ears of spaniels.

When, in cultural history, people evolved from good gatherers and hunters to farmers, shepherds, and city dwellers, they began to breed dogs selectively for hunting, herding, fighting, vermin catching, and draft work. By Roman times these dogs fell into readily identifiable types by size and function. They were not breeds as we now think of them, true to specifics of size, appearance, color, and behavior. Instead, they were dogs which shared useful characteristics—heavy, aggressive animals for guard or fighting duties; dogs almost as large but more placid and thus suitable for draft work, or animals whose scenting powers made them useful in hunting. Such types represented the extent of distinctions among European domestic dogs until about 150 years ago, when they were first bred systematically and more or less scientifically for what we now think of as specific breed characteristics.

If we can narrow the possible place of origin of the dog to either Europe or the Near East, and can date the first known dogs to about 6500 B.C., what can we say about the wild ancestor of the dog? This question has long been obscured by the extreme external variability of dogs. As recently as 50 years ago, it was still theorized that the small, foxy-looking breeds must be descended from foxes, the very large breeds from wolves, the American Indian dogs from coyotes, etc. We now know that we need not postulate more than one common ancestor—that all breeds of the domestic dogs are undeniably one species. There does, however, have to be a capacity for wide genetic variation on the part of this single ancestor.

The wolf and the golden jackal have always been considered

the two most likely ancestors. The golden jackal, after all, is found in northern Africa and southern Eurasia, just that part of the world in which dogs were early associated with men. The jackal is an animal of medium size, like early dogs, and it lives in proximity to man. At first glance, so compelling are the "common sense" arguments in favor of jackal ancestry that as noteworthy an animal authority as Konrad Lorenz until recently supported the theory of dual ancestry of the domestic dog, with most breeds stemming largely or completely from the golden jackal. But scientific fact makes jackal ancestry extremely unlikely. Dental characteristics of the jackal are distinctly unlike those of domestic dogs and wolves. Vocal patterns of dogs and jackals are likewise quite different, whereas dogs and wolves are very similar in voice patterns. (This consideration caused Lorenz to withdraw his theory of jackal ancestry.) Social patterns of the dog, a highly gregarious animal, do not resemble those of the jackal, which forms only pairs and temporary family groups, not packs.

Why has the gray wolf not been more easily accepted as the ancestor of man's best friend, if all information points most clearly to the wolf as the parent stock? For one thing, until very recently not enough was known about the wolf for people to recognize the great similarities between the two species. It was thought, for instance, that wolves did not bark; how was it possible, then, for the barking dog to have come from the howling wolf? Now we know that wolves do bark, and that man has selected barking as a desirable trait to reinforce. Anatomical evidence shows the very clear skeletal relationship between the wolf and less extreme breeds of dogs. The teeth of the two species are of the same type, those of the dog differing primarily in the reduction in size of the first molar or carnassial.

Behavioral evidence is also beginning to be useful in underlining the relationships of dogs and gray wolves. J. P. Scott and John L. Fuller, in their key work on the genetic basis of social behavior of dogs, described some 90 behavior patterns in the domestic dog and found all but 19 of them in the wolf. Most of those not recorded for wolves are minor patterns which probably have simply escaped the attention of field observers. The few patterns noted of wolves which the authors did not observe in their dogs are specialized types of hunting behavior unlikely to occur in dogs living in pens and yards. Scott and Fuller felt justified in concluding, on the basis of behavioral evidence, that the dog could only have been domesticated from the gray wolf.

The major problem with full acceptance of the wolf as the ancestor of the dog has been the matter of size. The northern gray wolf is far larger, with much heavier teeth, than the earliest known domestic dog, and it is stretching credibility a little far to consider it the dog's immediate forebear. But all along the southern borders of the wolf's range are subspecies considerably smaller than the very large northern animals. Such races as the Indian wolf are medium-sized and have smaller carnassials than the more northern races, and many authorities feel that the original dog came from a subspecies similar to these.

Although we do not know all the evolutionary steps that lie between the wolf and the domestic dog, one of the missing links seems to be the dingo-pariah group of semi-domesticated dogs. The dingo has its own chapter in this book, for in Australia it lives almost completely wild. The dingo appears, however, to be one of a large and widespread group of dogs called pariahs that are found across southern Asia and Europe as well as northern and eastern Africa. Pariahs have been known for thousands of years, and were once thought of as mongrels which had escaped domestication and taken to the wild. Modern researchers are inclined to think that they represent an original strain of wild dog which has not yet become fully domesticated.

Despite their immense geographical range, pariahs and dingo-like dogs show considerable similarity. They are animals of medium size, often tawny or reddish, though some are black or black and white. The coat is usually of medium length, often thick, though both short-haired and long-haired types occur. Pariahs commonly have prick ears that turn obliquely outward rather than stand perfectly straight. Within this general description several types occur according to geographical regions; in the Caucasus and Turkey, for instance, the pariah is usually quite large and heavy, with an especially long, thick coat, while in the Mediterranean a greyhound type appears. Several noted dog specialists have taken a special interest in the pariah group. Two Israeli specialists developed one type of pariah into a breed, the

Canaan Dog, which has been recognized by the Israel Kennel Club, accorded an official standard, and recently imported into the United States.

The constant characteristics of the pariah group, which have maintained themselves over millenia, seem to argue against these animals' being simply mongrels fleeing domestication. They appear originally to have been wild animals which have achieved and maintained a semi-domesticated, more or less symbiotic, relationship with man. To the dog the advantage of this arrangement lies in the food available around towns and villages, while the advantage to man is a steady garbage collection system and a network of alert guards. Some pariahs became sufficiently domesticated to be taken with man to the islands of the Pacific, as is the case with the dingo and the native dogs of New Guinea, New Ireland, and Polynesia, while in other places dogs indistinguishable from pariahs are used as guard and herd dogs.

The question remains: Is the domestic dog a descendant of the pariah-dingo group? Or are the pariah-dingos and the domestic dog both descended from an original wild dingo, very close to the wolf, which ranged the Near East? Or are the various dogs in question descended from a southern race of the gray wolf? At present we do not know. But this is a question which will surely yield, at least in part, to continued scientific examination. Eventually genetic, anatomical, archeological, and behavioral information on the entire genus should provide a clearer picture.

It is interesting to know what zoologists look for in distinguishing domestic from wild dogs, and in what ways the behavior and physique of the dog have been modified by domestication. The physical changes in the more extreme breeds of modern dogs when compared with the wolf are obvious to any layman. But physical characteristics of domestication show up in dogs that are superficially very wolf-like. The dog's back is usually much shorter and straighter than that of a wolf, and in cross section the chest is more barrel-shaped than the keeled chest of the wolf. This barrel-shaped chest results in a more or less turned-out elbow, exaggerated in many breeds, so that for most dogs the print of the hind foot falls beside that of the fore foot, while for most wolves the hind footprint either falls in that of the fore foot or lies in the same plane.

The eyes of domestic dogs are different from those of the wild members of the genus. Dogs have rounder eyes, directed more or less forward, while wolves' eyes slant slightly and often appear to be farther to the side of the head. Another modification of the dog's skull involves the raising of the brain case above the bridge of the nose, accompanied by an enlargement of the frontal sinuses and a broadening across the cheekbones. Thus many domestic dogs have a prominent "stop" below the eyes, sometimes almost perpendicular, between the almost horizontal planes of skull and muzzle.

Other bodily characteristics frequently occur in domestic dogs which distinguish them from wild species. Curled tails occur in many of the oldest types of domesticated dogs, greyhounds from Egypt and the near East, Chow types found in China and across northern Eurasia, American Indian dogs, and primitive breeds of African dogs such as the prototype of the Basenji. The curled tail, which no wild canid has, is not necessarily an indication of close relationship among the types involved. It is a mutation which seems to have been retained independently by various people in differing places, perhaps as nothing more than a convenient means of distinguishing their dogs from the wild ones of the area.

Mutant features like extremely short legs have also been retained independently in North Africa, Europe, and America; in some cases because a short-legged hunting dog was easier for the human hunter to follow on foot, or because such an animal could more conveniently go to ground after small mammals (as does the dachshund, developed for taking badgers). Another common feature among domestic dogs is the lop ear. Found only in pups of the rest of the dog family, the pendant ear is retained in many adult domestic dogs. For wolves a lop ear would be non-adaptive; it interferes with hearing. But it is no accident that most European breeds of hounds and bird dogs have lop ears; breeds which are supposed to concentrate on hunting by sight or scent would only be distracted by an unnecessarily acute sense of hearing.

None of these common marks of domestication occurs in all dogs. Although recent techniques of multicharacter analysis of skulls allow highly trained zoologists to use a whole series of proportional measurements to distinguish various species, the single most consistent and useful means of separating the domestic dog from the wild one is by the teeth. The dog's teeth, especially the canines and carnassials, are on the whole smaller than those of the wolf. In the dog the carnassial of the upper jaw is usually shorter than the combined length of the two molars which follow it, while in most races of the gray wolf it is longer. This difference in size reflects the behavior of the animals: the wolf has the largest carnassials among the genus because it is the most highly carnivorous and the carnassials are the ripping teeth used to tear apart large chunks of meat.

Does this mean that the teeth of the domestic dog have regressed, so to speak, have somehow returned to a more primitive, less specialized form? To some extent this may be possible; the skeletons of young mammals have a considerable degree of plasticity and the type of life a young animal leads has some effect on the growth rates of various parts of the skull. Possibly, also, man has selected dogs with less formidable teeth. Most likely the smaller size and proportions of the teeth of domestic dogs reflect the skull and teeth proportions of the ancestral type, such as a southern race of the gray wolf.

More important to man than the physical changes brought about by domestication are the behavioral changes. The overall effect of domestication has been to make dogs far more various in behavior than they could be in the wild. The wild dog must be a well-rounded individual to survive. Even among pack animals (where some variety can be tolerated) each wolf must be able to run fast enough, hear, see, and smell well enough, think and react quickly enough, to stay out of danger and to feed itself. But for the domestic animal survival depends on pleasing the master, on doing what *he* wants done.

Some behavioral differences between wolves and certain dogs can, of course, be explained by physical peculiarities. Short-legged breeds cannot run very fast; very heavy breeds lack agility; breeds with hair in their eyes have trouble seeing; breeds with bulldog noses frequently have difficulty breathing. What would be crippling disadvantages to a wolf, however, are offset by capacities which human beings regard as advantages. The bulldog with its deformed snout can take and maintain a nosehold on a bull or other large prey. A very heavy stocky breed may make up in pulling power what it lacks in agility. Dogs with hairy faces depend on scent and hearing to perform their tasks.

But there are some behavioral changes which do not have anything so obviously to do with physical peculiarities. Scott and Fuller found that while basic behavior in dogs and wolves is essentially the same, there are some significant differences in patterns of investigation, hostility, and sexual behavior. Changes in sexual behavior in the domestic dog have led to increased fecundity. Wolves breed once a year, females coming in heat first at two years of age. The average litter size is four or five pups. This number of young is all that even the most fortunate parents can hope to raise in a year. More pups would be a disadvantage to the pack, and would be unlikely to live to breeding age themselves. But when man helps provide for the pups, a large litter is as efficient as a small one, and the more frequently a bitch produces pups the more valuable she is. So man has selectively bred females which mature young and bear large litters. Litter size among domestic dogs varies considerably, larger breeds not uncommonly having eight, ten, or 12 pups—the highest recorded number is 23. Toy breeds, on the other hand, often average only from one to three pups, with still-births common and postnatal mortality high; obviously an extreme reduction in size is genetically and physiologically unfavorable. The age of first heat also varies with the breed, small breeds coming in heat initially at one year or even younger, large breeds often at around 18 months.

An even more significant difference between the sexual behavior of domestic and wild dogs is that most domestic bitches have two fertile periods a year, while wolves have only one. Dingos and pariahs, and among domestic dogs the Basenji, also have a single yearly cycle. In domestic dogs the hormones of the ovaries and pituitary which control the sexual cycle act independently of external stimulation, but in wolves, dingos, and Basenjis the pituitary must be stimulated by light changes to trigger a new cycle. By crossing Basenjis with other breeds of dogs it has been discovered that this dependence on light changes is the result of the actions of a recessive gene.

Wolves show hostile behavior towards prey and towards other

wolves. Dogs have much the same patterns of attacking prey as wolves, working singly or in groups and sometimes using a herding or other cooperative maneuver. But in some breeds hunting behavior has been restrained so that the attack is not carried to the death. Herding dogs, for instance, must show enough hostility towards sheep or cattle to get them to move, but not enough to hurt them. Bird dogs must be sufficiently interested in hunting birds to locate them, but must be able to be controlled from a distance by the human hunter. Retrievers must be eager to find and seize the prey but their bite is inhibited (the dog's mouth is "soft") so that the retrieved game is not mangled, much less eaten.

Techniques of fighting with each other are much the same in dogs and wolves, though there may be some variation in tactics among different domestic breeds. In many, however, human intervention has modified the inclination to fight. Terriers have long been bred for their scrappiness, but in scent hounds and bird dogs, for instance, the urge to fight has been greatly reduced. These dogs are used for hunting in packs, or are kept in large kennels; it is therefore sensible to select animals that have no interest in fighting other members of the group, and that keep their attentions strictly on hunting. Thus, though all dogs fight much the same way when aroused, some breeds fight readily and others can hardly be stimulated to fight at all.

Investigatory behavior in dogs and wolves may either be social, or it may involve investigation of the environment when hunting. Social investigation is much the same in the two species, attention being directed to the anal and genital regions and to the face. In hunting behavior, wolves are generalists; they use sight, hearing, and smell—and all three of these senses must be reasonably sharp. Various of the dog breeds have become specialized, however. The coursing hounds run their prey by sight; the scent hounds rely almost exclusively on their noses. Bird dogs use both eyes and nose and retrievers mark falling birds by sight.

The most important change in the behavior of such specialized breeds of dogs seems to be the strengthening or reduction of the effects of certain kinds of stimulation. Shepherd breeds are highly stimulated by the smell of sheep or even deer, and bird dogs by the smell of birds. Many terriers, on the other hand, have little interest in scent. Scott and Fuller conducted an experiment which graphically demonstrates differences in the scent curiosity of different breeds. They put a live mouse in a one-acre field and then let in a group of beagles. The beagles found the mouse in less than a minute. A group of fox terriers took 15 minutes to locate their mouse, while Scottish terriers never did succeed in finding theirs (though one of them stepped on the mouse without noticing it). The results do not indicate that terriers are either stupid or deficient in scenting ability. They seem simply to be uninterested. Fox terriers, for instance, find their stimulation in sounds.

Man's selection, then, has not conspicuously changed the physical capacity of various breeds of dogs to use their eyes, ears, and noses (except in obvious cases where physical characteristics may affect the sense involved). But it has served to change the level at which a breed is stimulated by various sensory messages, and in many cases to concentrate the attention of certain animals on messages from one sense alone.

Semyen Fox

(Simenia simensis)

IN THE HIGH PLATEAUS and mountains of Ethiopia lives a mysterious wild dog about which so little is known that even its correct scientific status is still in question. Its myriad common names reveal our ignorance: it has been called the Semyen, Abyssinian or Ethiopian red fox, red jackal, red dog or wolf—in short, it has been linked with all the familiar kinds of wild canids. The name Semyen fox is most common; the animal does live on the Semyen plateau of north central Ethiopia and is fox-like in voice and behavior, though it is certainly not closely related to the vulpine foxes.

The only other locality from which the Semyen fox has been reported is the southcentral part of the country, in the Bale mountains. It has never been found outside Ethiopia. Both areas in which it occurs are high plateaus, rising to over 10,000 feet and covered with moorland. Semyen and the Bale mountains are separated from each other by several hundred miles of lower country in which the fox is not found, and it is not surprising that the animals found in the two regious differ slightly in color and size.

Within the past few years the Ethiopian government has embarked upon creation of the Semyen Mountains National Park, which includes the most remote portion of the Semyen plateau and Ethiopia's highest peak, Ras Dashan. The activities connected with setting up the Park have brought several interested observers in contact with the Semyen fox, so that bits of information and disconnected observations are now available on the animal.

The Semyen fox is red, the Semyen plateau race being a reddish-yellow and the Bale race, slightly larger, a darker red. The tail is blackish with a white tip. The belly, insides of the legs, feet, and throat are a light buff or white. There is a line of white bordered by dark along the side of the neck. Though its pointed face and large triangular ears give it a foxy look, the Semyen fox stands about 24 inches high at the shoulder, far taller than any fox, and is conspicuously long-legged. Its voice is a high-pitched, drawn-out, screaming bark, which may be repeated for several minutes on end. This call is heard particularly at dawn and dusk. The animal has also been heard to yap sharply when moving away from a person who has surprised it.

Although little is known about its living habits, what has been reported is remarkably consistent. It is seen most often at dawn or at dusk, and therefore might be assumed to be nocturnal. However, it is observed occasionally during the day in areas where it is undisturbed and its nocturnal habits may simply be an adaptation to the increasing encroachment of man on its habitat. It is not particularly shy—only cautious, and if surprised by man will withdraw, sometimes watching curiously from a distance before making off. Seen singly or in pairs, it has also been sighted in groups of four or five, apparently adults with half-grown pups.

The Semyen fox appears to live in a den among the rocks during the day, and may inhabit the same den for a long time. One observer watched a solitary male several times from a blind near its den, and reports that it was slothful, given to extended periods of barking, and never left the den area while he was watching. The grass at the mouth of a den is flattened, and the animal may scrape out a bed on the floor if using one of the caves common to the rocky crags of the Ethiopian uplands.

The moors in which the fox is found are densely populated by two species of grass rats which form the staple of the fox's diet. In some places these rats are so common that their squeaks are plainly heard as one walks through the grass tufts. Where rats are thick, foxes are common. They hunt along paths and among the tussocks, apparently catching the rats above ground rather than digging for them. The fox leaves frequent droppings, composed

Semyen Fox in its moorland habitat in the mountains of Ethopia. Despite its name it looks more like a coyote or jackal than a fox. Picture at right shows its long legs and handsome tail markings. *Photo: M. Bolton*

almost exclusively of rat fur, teeth, and bones. Observers agree that rats form the bulk of a fox's diet, though it may well eat other rodents and perhaps such insects as locusts.

Local people accuse the Semyen fox of preying on domestic lambs and kids, which it almost certainly does not do with any regularity. To begin with, the fox was in this area long before the recently-arriving Ethiopian farmers and herders began to bring their flocks into the high plateaus, so it evolved independent of man and his stock and adapted to a rodent diet. Nor has anyone found a shred of evidence that it preys on sheep; its habits are not those of a predator on larger animals, and its droppings show no signs of wool or livestock remains. Although such negative evidence does not preclude its occasionally eating a lamb or a kid, the fox does not commonly do so.

The Semyen fox apparently has few natural enemies. Although it sometimes comes down into the forests of giant heath it is primarily an animal of the highland moors, above the range of the hyena and leopard. Fox pups may fall victim to the many large birds of prey found in the Ethiopian highlands, since eagles are known to take pups of other wild dogs. Though natives prefer to save their scarce bullets for edible game like the nyala and the rapidly disappearing Walia ibex, some foxes are shot. In Semyen, at least, the fox seems to be subject to rabies or a similar disease which causes it to act strange, to run amuck through villages, and eventually to die. Whatever the disease, it apparently has been partly responsible for the fact that the fox is becoming scarce in the Ras Dashan area.

The people of Semyen endanger the fox primarily by their destruction of its habitat. The animal has evolved in an alpine environment relatively free from predators and abundant in its favored food of grass rats. But man is moving into the area, farming as high as he can on the mountain slopes, gradually stripping the cover from the land and eroding the soil. He is also bringing domestic dogs, which will present a grave hazard to the fox. By 1967 observers reported that on the Semyen plateau the fox was being seen more and more rarely (though in Bale it was still quite common), and that it was probably suffering from a combination of disease and pressure from man. It would be unfortunate indeed if this creature disappeared even before man knew what he was losing.

The Genus Vulpes

THE FOXES of the genus *Vulpes* make up the largest and widest-ranging group of wild dogs. Its nine species are found almost everywhere in the northern temperate zone and in much of Africa. Thanks to man, it is also well established in Australia. Vulpine foxes are adaptable to a great range of conditions. Some species are distinctly desert animals. Others inhabit open plains and semi-desert or very high mountain plateaus. The most widespread and well-known member, the red fox, lives on farm and woodland and thrives in close proximity to agricultural and suburban man.

The vulpine foxes are quite similar in size and general appearance. All are rather small, ranging from four or five to 14 or 15 pounds. They have erect pointed ears, a slender, "foxy" muzzle, and a distinctive tail. This tail, or "brush," is long, straight, and heavily haired all around, and extends behind rather than being pressed against the hind quarters or raised above the back as in *Canis.* It lacks the expressiveness of the tail of the wolf or domestic dog; it may be wagged or lashed stiffly, but it cannot be curled at all.

Vulpine foxes also share a distinctive feature of the eye. When the pupil is exposed to strong light it forms an ellipsis, a vertical slit, rather than becoming small and round as in *Canis* (or in human beings.) All dogs have scent glands on the soles of the feet, around the anus, and on the top of the tail at the base, but in some members of the genus *Vulpes,* notably the red fox, these glands are especially large and produce a scent which is noticeable even to man. Members of the genus have from four to eight mammae, but typically six; like most other dogs, they have 42 teeth.

Unlike the large pack-hunting dogs, foxes do not need to be social. Their normal prey are rodents and rabbits, which they can easily kill by themselves. Red foxes are solitary except during the breeding season, when adults associate with their mates and, for a short time, with the young. For the most part red foxes den by themselves, but other vulpine foxes, such as the corsac, may den in groups or colonies. All members of the genus have a gestation period of about 52 days, but the length of time pups stay at the burrow and in contact with the parents varies. The red fox family breaks up about four months after the birth of the pups, while kit fox puppies of six or seven months old may still live at the parental burrow. All vulpine foxes are mature at a year.

North America has two representatives of the genus *Vulpes.* The red fox, found as far south as the northern border of Mexico, was once considered distinctly American but now is known to be the same species as the Eurasian red fox. The changing habitat of the United States has allowed it to thrive and to extend its range. The other North American representative is the kit fox, once found throughout the great plains of the United States and Canada and across the deserts of the American southwest. But the poisoning campaigns of the 1800's, carried on primarily by hunters of wolf and coyote pelts, succeeded in virtually exterminating this pretty, unwary little animal from the plains. Fortunately the kit fox survives and is staging a comeback in portions of western deserts where few hunters go.

Eurasia has several representatives of the genus *Vulpes.* The red fox exists in greater numbers in many portions of agricultural Europe and Asia than it did in the days of the primeval forests, which were far less attractive to it than the present mixed habitat of field and woodland. The steppes and deserts of central Asia are the home of the corsac fox, a more social animal much hunted in the Soviet Union for its pelt. Also in central Asia, from the eastern edge of the Caspian Sea to the mountains of West

Pakistan, lives the small, obscure hoary fox, an animal seemingly well adapted to a severely arid climate. Another fox well suited to severe conditions is the Tibetan sand fox, which hunts the rodents of the high windswept plateaus of Tibet at an elevation above 13,000 feet. The other exclusively Asian fox is the Bengal—about the size of a small red fox—which is found throughout the Indian subcontinent in open country and scrubland, not heavy jungle, and is common around villages and inhabited areas.

One of the most highly desert-adapted of the vulpine foxes is Rüppell's sand fox, a lovely little animal that occurs from Afghanistan through the Near East and across North Africa. Rüppell's fox has the very long ears of desert canids, and carries a strongly marked white tail tip which helps distinguish it from the other small desert foxes. It is the most northern representative of the genus in Africa. In the band of savanna which stretches east and west aross the continent between the Saharan wastes to the north and the tropical vegetation of central Africa is found another fox adapted to semi-arid conditions, this one called the African sand fox. This sand fox is, if anything, more obscure than Rüppell's, though it has been successfully kept and bred in captivity. Finally, the southern temperate zone of Africa is the home of the Cape fox, another small, long-eared animal that ranges the open plains and velds, feeding on rodents and other small mammals.

The chief importance of the vulpine foxes to man lies in their integral place in the ecology of so many regions. While no predators, not even foxes, can control irruptions of rodents, under normal conditions they play their part in keeping small mammal populations at a healthy size. As a hunter, man has always regarded the fox as excellent sport. The red fox has built a world-wide reputation on its ability to "out-fox" the hunter, whether he's mounted or afoot, alone or accompanied by highly trained dogs. The British, finding the red fox absent from much of the Indian subcontinent, gladly took up the pursuit of the Bengal fox, which while reputedly not quite such good sport was still a highly acceptable diversion.

The red fox has also been valuable to man for its fur. Red foxes exist in the wild in many different and beautiful color phases and the most valuable of these have been very successfully bred on fox farms. To modern urbanized man, however, the esthetic value of foxes may well be their greatest contribution, for these beautiful little predators bring the essence of wildness to our very backyards.

THE RED FOX (*Vulpes vulpes*)

To people who live within its range, the red fox needs little introduction. It is the cunning and resourceful Renard of myths and stories; it is the animal whose pelt, in its many beautiful color variations, has been one of the most valuable and sought-after of furs; it is the object of pursuit by that curious Anglo-Saxon cult which hunts to hounds. For all its familiarity, however, the red fox bears the burden of much popular and even scientific misinformation. Though they have long been bred by the fur trade, relatively little is known about the lives of wild foxes, basically because they are difficult animals to get to know. Red foxes are solitary and mobile, for much of the year without a permanent den. They are small, alert animals with very sharp senses whose stock in trade is to avoid drawing attention to themselves. So the familiar red fox is not so familiar after all

The red fox has an immense range roughly similar to the original scope of the gray wolf. In Europe it extends to the very north, except for some of the northern islands, while in Asia it stops for the most part at the northern tree line. South, it is found through Europe and into North Africa north of the Sahara, in the Near East, and through temperate Asia into northwest India and northern Vietnam. It occurs in Japan, Sakhalin, and the Kuriles but not on the more southern Pacific islands. It is an introduced species in Australia. In America it is found from the Arctic to the extreme southern United States.

The introduction of red foxes into areas where they did not originally occur was the work of British sportsmen dedicated to fox-hunting. Imported to Australia in the late 19th century, the red fox is now found throughout much of the continent. Settling into a virtually unoccupied ecological niche, the fox may have hastened the decline of the native Australian predators. Hunted, trapped, poisoned, and bountied in Australia with no apparent effect on its numbers, the red fox has seemed impervious to pressure from man so long as the habitat remains favorable.

In North American it is both a native animal and an import. Bone remains from aboriginal sites in eastern North America indicate that the red fox occurred before the arrival of Europeans in southern Ontario. Pre-European sites farther south, notably in southern Pennsylvania, bear no evidence of red fox remains, though remains of the gray fox as well as a complete range of other native mammals are common.

Historical evidence supports archeological finds. The earliest settlers in the northern Atlantic states noted the presence of the gray fox, but the red was found only in northern New England, was quite rare, and was restricted to black-phase animals. Southern New England and the middle Atlantic seaboard, judging from historical records, had no red foxes. Evidently in pre-European North America the red fox was restricted to Alaska, Canada, and the northern portion of New England.

But British settlers could hardly be expected to do without the pleasures of fox-hunting, for which the gray fox was not at all suited, since he was inclined to den up or take to the trees rather than give the hounds a good run. The importation of red foxes is first recorded in about 1650, when several pairs brought from England were released along the eastern shore of Maryland. In the early 1700s imported foxes were turned loose on Long Island (from which they reached the mainland of New York State over winter ice). Imports were also made to Virginia and New Jersey.

This introduction of red foxes from Europe coincided with the clearing of the forests and the development of farms, changing the habitat which had previously been inhospitable to the species. The imports flourished in a newly rural environment and it is possible that the native red fox from farther north was encouraged to move south. Certainly imported stock followed settlers north, so that the red foxes now found in eastern North America must be a complete mixture of imported and native strains, with the imported strain predominating. The red fox has been expanding its range in the United States ever since this first introduction and is now found throughout the country. It has also pushed north as far as Baffin and Devon Islands in the Arctic Ocean.

The red fox is a small animal whose long luxurious coat makes it seem larger and heavier than it is. It varies considerably in size from one area to another, more southern animals tending to be smaller and lighter than those living farther north, and animals from the mountains generally being larger than those from the lowlands. Within a given area, females are usually smaller than males. The body length of the red fox varies from as little as 20 inches to as much as 35 inches. The tail can be from 13 to 18 inches long. In the U.S. the red fox usually stands about 15 inches at the shoulder.

It weighs from eight to 12 lbs., with males averaging around 11½ and females nine or ten. (A small fox of one of the southern races may weigh five or six lbs., while large northern males sometimes exceed 20 lbs.). There is considerable seasonal variation in weight, since animals are generally heaviest in mid-winter and lightest in the spring—when the male is recovering from the rigors of the breeding season and the female is both suckling and hunting for the pups.

Despite the common name, color in the red fox is distinctly variable. Typically it is a vivid red over the back and head, with white chest and belly, black on the backs of the ears and on the feet, and a white tip to the tail. (This white tail tip is not the invariable identifying feature of the red fox it is often claimed to be, however; sometimes the tail tip is black, a distinctly yellowish buff, or dark red.)

Although the pattern of vivid reddish back and white underparts is the most common, the red fox exists in the wild in three genetic color phases, red, black (or silver), and cross. But mention of these three main color types does not begin to describe all the lesser color variations. The red phase usually has white underparts, but sometimes animals with red backs have dark gray or black bellies. Some animals of the red phase, especially along the northern seacoasts or in the Arctic tundra, are a very light yellow or buff instead of deep red. In the cross phase the barely discernable band of darker hair which is present in most red foxes is accentuated; although the rest of the coat is a dark red, there is a prominent black band which runs across the shoulders and extends down the backbone to the tail.

The black phase fox may be solid black, brownish-black, blue, or silver, the latter caused by the white tips of the guard hairs which stand out against the dark background. Among the bluish members of this phase can be the blue cross fox, in which the dark cross pattern on the back is prominent enough to show even against the bluish coat. Albino foxes occur occasionally, as do partial albinos, which the Russians call "ermine" foxes. There are also gray foxes, but it is doubtful that gray is a color phase as such. Adult foxes may have gray patches on the hips, but evidence indicates that grayness over the rest of the body is the result of age. Various forms of the black mutant, as well as the cross fox, are commercially more valuable than the ordinary red phase, and silver and blue foxes have been bred extensively on fur farms, although the market for fox pelts has dropped considerably.

The red fox molts once a year in the spring and the process is completed by fall. As molt begins, the long guard hairs become coarse and brittle, lose their luster, and begin to fall, first from the head, shoulders and flanks, and finally from the rump. As molt continues the thick underfur comes out in clumps, which leaves the fox with an unkempt look, but by summer the underfur is nearly gone, and the new guard hairs lie sleek against the body. In late summer the underfur begins to grow back and gradually thickens until by winter it is dense enough to cause the guard hairs to stand out from the body and make the animal look much heavier and more rounded than it had a few months before. Young foxes, born with only the woolly undercoat, grow long guard hairs over the summer to complete their first winter coats, and molt for the first time in their second summer.

The scent glands are particularly well developed in foxes. On each foot there is a gland, visible when the foot is spread, which opens in front of the center pad. The scent from this gland leaves a well defined foot trail the fox can follow if it wishes to double back in its tracks, and which can also, of course, be followed by predators. Another scent gland, on the upper side of the tail at the base, is usually plainly marked by a small patch of black fur. The glands below the tail on either side of the anus produce the pungent scent, typical of vulpine foxes, which is conspicuous even to humans. These glands, used deliberately for scent-marking objects, also emit scent involuntarily when the animal is frightened. (Foxhunters tell us that a closely pursued animal produces an especially plain scent trail.)

All this odor, and especially the tendency to produce more scent when pursued, is certainly not helpful to a prey animal, which the red fox, thanks to man, has become. But we must remember that before the hunger and his dogs, the fox had no systematic predators. The fact that excessive scent production has not been eliminated by natural selection is a pretty good indication, with others, that the fox was never hunted to any degree by the wolf or other natural enemies. The laying of a strong scent trail is probably a social phenomenon; originally the only animals likely to be interested in a fox trail were other foxes; rivals of the same sex were thus warned to avoid each other, and during the breeding season foxes of opposite sex used scent to seek each other. Man now uses the fox's scent-production to its disadvantage, and the fox has not had enough evolutionary time to eliminate this characteristic.

The footprints of the red fox are slender and longer than they are wide. Each foot shows four long claw marks. The hind foot, on which the center claws sometimes converge or even cross, is slightly smaller than the fore foot. A clear print in mud or snow often shows the mark of the heel bar on the center pad and sometimes the impression left by the hairs between the pads. As the fox trots it places its feet in an almost straight line, with the hind feet falling in the marks left by the fore feet, so that there is a distinctive line of what appear to be single prints left by a clawed pogo stick. On snow or very wet ground a brush mark is sometimes left by the fox's trailing tail.

Normally the fox is hunting as it moves, so that a typical trail shows a multitude of little side excursions as the fox investigates sounds or smells, and often there are signs where the animal has stopped to sniff, dig, or pounce at possible prey. A trotting fox covers about five miles an hour, but he can gallop as fast as 35 miles an hour.

Foxes are basically solitary animals for which communication is as much a means of keeping individuals apart as bringing them together. Although they are usually rather silent, foxes have a wide variety of calls—they bark, scream, howl, yap, growl, and even seem to hiccup. During the breeding season various sounds serve both to bring males and females together and to warn off interlopers. In the winter, the male fox is attempting to establish a territory and to mate with the female or females within that territory. During this time of the year a yelling bark, either short and broken or a long single note, is heard, usually in the evening and often from an animal moving in a regular pattern. Such a "wo-wo-wo" bark seems to be made by a male fox to indicate the territory he is staking out and to warn other males not to intrude.

A contact bark is a softer sound made during the breeding season when the mated pair are traveling together or close to one another. A third barking sound is a light yap, rather like the voice of a small terrier, which carries only a short distance. With all these sounds there is such variation among individuals that a person can learn to distinguish one animal's bark from another's by its relative harshness, pitch, duration, and so on.

Foxes also scream, a hair-raising screech of varying duration which sounds like nothing so much as cat-fighting. Tradition has long supposed the noise to come only from vixens, but actually the scream is made by both sexes and by half-grown pups as well. It seems to be aggressive; pups scream at each other during their play-fighting, and adults of the same or opposite sex will put their faces close together and keep up the noise as a form of bloodless combat for as much as 10 or 15 minutes. The performance also seems, however, to have something to do with mating; screaming animals have been filmed before and during copulation attempts.

Foxes can also produce a piercing howl, occasionally repeated several times. Heard at various times of the year, the howl may serve as a warning to pups. Other adult sounds directed to the young include a soft growl followed by a series of hiccupping noises which warn of immediate danger in the area, such as the close presence of man, and a soft murmuring sound made to call the pups from the den.

Important as the voice is in fox communication, it is probably less important than scent. Foxes mark their living areas by the use of feces, urine, and scent from the subcaudal glands. At any time of the year the presence of droppings and urine inform foxes of each other's location and condition, but scenting is most important before and during the breeding season. Fox excreta is rather like that produced by domestic dogs, but when fresh it has a foxy odor and frequently is drawn out at one or both ends into a tail formed by the prey hair or grass it contains.

Scent marking with feces is most common in the late fall and winter, when droppings are concentrated in certain parts of the territory. Gaps in hedges are a common point of concentration, as are the edges of fields or any conspicuous point of passage in the animal's regular trail. A cluster of droppings will be found at one gap or point, and then perhaps a week later another point will be more heavily marked. At all times of the year, deposits are regularly made on certain objects—bird and other prey remains, old tree stumps, cow dung, bare or burned ground—which foxes visit regularly.

As with domestic dogs, foxes use urine to mark their territories, especially during the breeding season. They also rub their subcaudal glands against trees, bushes and clumps of grass, walking stiffly and with arched tail from spot to spot.

Foxes, like dogs in general, see quite poorly, though they spot movement readily. Their eyes give them a picture of their surroundings so that they are able to perceive objects which have been added or removed, but they seem not to be able to pick out stationary objects distinctly. The fox does have good night vision, far better in comparison with man's than its day vision, and it thus can see movement at night which would be unnoticed by man, though of course it cannot distinguish stationary objects in the dark any better than in the daytime. When the fox hunts, it depends on its ears and nose for information about food.

The animal normally travels at a trot, though it sometimes ambles slowly. At a gallop its agility and adeptness at dodging and doubling back make it difficult for any pursuer to catch. The red fox can climb inclined branches of trees, and scramble onto walls, so that resting foxes are sometimes found in the crotches of trees a few feet above the ground, or on low walls or wood piles. But since the red fox, unlike the American gray, cannot actually scale trees vertically it can hardly be credited with real climbing ability.

Red foxes are good swimmers, most sources agree, but they seem to avoid water. In areas where there are small streams and rivulets a common feature of fox trails is the water jump, a point where resident foxes regularly cross a stream by jumping, sometimes followed by a scramble up a steep bank. Fox hunters credit their prey with remarkable cunning in attempting to throw off its hound pursuers by crossing water. Actually the fox is merely using its regular trails and habits, just as it does when it clambers up a stone wall and runs along it before jumping off on the other wide, thus eluding dogs.

Foxes are nocturnal for most of the year. They leave their dens or bedding places in the evening, usually before nightfall, and return to some sort of concealment after sunrise. An exception to this pattern occurs during the mating season, when males are quite mobile in their search for territory and mates and travel more frequently and freely during the day than they do at other times. Another exception occurs when the female is feeding pups; then she is under heavy pressure to find food and sometimes hunts well into the day, or returns to the den in mid-day to suckle or bring food to the pups. Obviously, when the fox has little to fear from man it tends to range more freely in the daytime than it does on farmland or in suburban areas. Once pups are old enough to emerge from the den they often spend much of the day playing around its entrance, thus providing the easiest and most enjoyable fox-watching.

Adult foxes do not inhabit dens the year around. The female is proprietor; the den is the center of her home range most of the year. She stays in it in the daytime during the breeding season, and bears her young there. She is bound to the den for as long as the pups use it, though as they grow up she prefers to bed down at a distance to avoid their attentions, and returns less and less frequently as they are weaned and become independent. Once the pups have left, she may spend the late summer and early autumn mostly above ground, but by mid-autumn she moves back to the den to claim it as her own and begin preparations for another breeding season. The male fox, on the other hand, usually beds above ground throughout the year.

Bedding places can be located anywhere that offers concealment, protection from the weather, and more than one avenue of escape. Favored spots are overgrown ditches, thick hedges, tangled tree roots, or long grass. During the winter a fox likes to sleep in the sun where it can soak up warmth and conserve energy. Pups of both sexes, once they leave the home den, use temporary bedding places during their first year, until maturation causes the young vixens to search out a den and sends the young males on rambles in pursuit of territory and mates. A fox bedded down in the daytime usually cat-naps, sleeping for from 15 seconds to half a minute only to wake up, gaze around and then sleep again. But when a fox sleeps this way its senses are awake, as its twitching ears prove. It may start from sleep in response to a mouse squeak or the sounds of a ground squirrel many feet away. It sleeps soundly or heavily only when well hidden in heavy cover, and then will awaken only every hour or so to check its surroundings.

Features of the terrain and availability of food determine the size of a fox's hunting range; the winter range is often larger than the summer one, of course, because of relative food scarcity. Two adult males in Illinois which were tracked over a period of several months by means of radio transmitters fastened to collars were found to do their hunting within an area of approximately 1000 acres.

Most of us know foxes as either hunters or the hunted. The popular image is of an animal of great guile and stealth: a reddish killer stalking its prey. Actually the hunting fox is an opportunist, roaming widely and for the most part picking up what it happens across, rather than engaging in elaborate stalks or chases. As hunters, foxes are solitary. Though they may occasionally be found in numbers around carrion or dumps, they never hunt in packs.

Although small mammals, and occasionally poultry, are an

important part of the its diet, the fox consumes a great quantity of food which hardly fits its popular image—slugs and earthworms, reptiles and amphibians, insects of all sorts, and much fruit and vegetable matter. The fox is also a scavenger both of carrion and of human refuse. Much remains to be learned about the fox diet, especially about variations in different parts of their immense range. From what is known, however, some factors seem constant: the basic importance of small mammals, the wide use of fruit and vegetable matter (especially at certain times of the year), and the importance of carrion and refuse.

In central Alaska, when mice are abundant, they are the chief rodent prey of foxes and are pieced out in the summer by the hoards of ground squirrels which hibernate during the winter. Alaskan foxes seem to have no trouble catching snowshoe hares, and snap up an occasional ptarmigan. During the winter, carrion of moose, mountain sheep, and caribou is especially welcome. Blueberries and crowberries are found in many places during the summer, and in winter enough remain under the snow to provide seasonal variety.

In Pakistan and northwest India, half a world away, red foxes are as omnivorous as their Alaskan counterparts. They depend primarily on small mammals such as rats, squirrels, marmots, pikas, and hares, eat birds like francolins and pheasants, and fill out their diet with fruits, berries, insects and wild honey, picking up scraps around human dwellings when hunting is poor.

Throughout the red fox's range mice seem to be a staple in the diet. In the Soviet Union field mice and voles are the small mammals most often caught, as are voles in England. Before myxomatosis decimated the rabbit populations of Europe and Britain, rabbits constituted about 50 percent of the red fox diet in Britain. In the U.S., the ubiquitous cottontail rabbit is a staple item throughout the year, but is especially important during winter, when insect and most vegetable food disappears and meadow mice may be hidden under snow. The importance of different species of hares, which are bigger and faster than rabbits, varies from place to place.

Foxes hunt mice and voles in several ways. During the summer a fox can locate mice by smell and sound in their grass houses or runways, and can dig them out of their holes. When the ground is snow-covered, the fox plunges into the snow above them, hoping to pin them down with its paws. Even if it misses on the first dive, the pounce blocks the runway and keeps the rodents from escaping. Foxes do not try to hide when they are hunting mice; they simply wander, pouncing and digging, through good mouse territory. A fox can hear a mouse squeak 150 feet away if conditions are right; this fact is put to use by hunters who call foxes within shooting range by imitating mouse noises.

Foxes seem to chew even such small prey before swallowing it—all but feet and tail. Five mice a day is the estimated average, but there is on record one fox stomach that contained between 40 and 60 voles.

Besides mice and rats, rabbits and hares, foxes take a variety of other small mammals. Tree squirrels are hard to catch, even in places where they spend time on the ground, and seem to be eaten only occasionally. On the other hand, ground squirrels are accessible; in Alaska, they make up half of the fox's summer diet. Despite their availability, insectivores—moles, shrews, and hedgehogs—are not favorite fox foods; they are often killed but left. In feeding tests with captive foxes, shrews are generally rejected; many other predators react to shrews the same way—the taste is apparently just unpleasant. Moles are not eaten much either. However, foxes do eat hedgehogs which are abundant in Britain and on the continent. A fox tears away at the soft underparts and then eats out the flesh, leaving a cleaned skin behind. A hedgehog rolls into a ball to present its prickles to attacking predators, and a fox will sometimes push one into shallow water, forcing it to unroll and expose its stomach.

Birds are an occasional item in the fox diet, especially in the spring when they are nesting and foxes are feeding their pups. However, most birds are taken as a result of opportunism and nowhere do they form the mainstay of the fox's diet. The hunting fox is always alert for groundbuilt nests and nesting, young, or injured birds. With the increase in the use of toxic chemicals to dress seeds, birds which have died as a result of feeding on such seeds attract foxes, which in turn may die from eating them. Despite the occasional hysterics of bird hunters and sporting groups, however, fox predation on game birds, whether in Britain, the United States, or Australia, is lower than might be expected and seems to have little or no effect on their numbers.

Occasionally unusual circumstances allow foxes to kill large numbers of nesting birds; on a night of driving rain and heavy overcast foxes killed over 200 black-headed gulls in a breeding colony on the west coast of England. When large kills occur they are often in places where birds would not nest widely unless encouraged by man. For example, many shorebirds nest under natural conditions on fox-free islands rather than on mainland coastal areas. Some birds also have a group defense against foxes and other predators, the mechanism called "mobbing." Geese, for instance, gather from far and wide to mob a fox, trumpeting in great excitement, and even following the predator as he escapes. The mobbed fox seems far too uncomfortable in such a situation to try to catch his tormentors.

Vegetable matter, especially fruit, is of unexpected importance to an animal which we tend to think of as carnivorous. Foxes like fruit and eat it by preference when it is available, partly perhaps because it is easy to get. In North America wild blueberries are a special favorite, but foxes also feed heavily on apples, pears, and plums—and do not restrict themselves to the growing season. Much tree-growing fruit falls to the ground before it can be picked and stays in the long grass of orchards. Berries stay on the bushes and ground long after they have ripened. In orchard areas the presence of fruit during the autumn, when young foxes are feeding themselves, may be important in reducing pup mortality.

Besides fruit, foxes eat grain, silage, acorns, clover, moss, and especially grass. So much grass is sometimes found in fox excreta that it appears foxes eat it deliberately. (On the other hand, in pup scats the inevitable presence of small amounts of grass mixed with insect remains indicates that young foxes have more trouble than their elders in separating animal prey from vegetation.)

Yet another source of fox food is the carrion provided by other predators. Under natural conditions the great carrion-producer for the fox was the gray wolf. In places where the wolf still exists wolf-killed remains may give the fox the extra measure of security it needs to survive year round. On Isle Royale in Lake Superior, for instance, many sources of food are scarce in winter, but the foxes thrive then because they profit from moose kills made by wolves.

Other predators are not the main source of carrion for the

fox, however. Man is not only the most destructive hunter, but the one least likely to use the animals he kills or wounds. An upland bird or waterfowl is often wounded only to be lost among dense brush or swamp; a deer hunter too frequently shoots at anything that moves, only to depart rapidly if he has killed an illegal doe or a farmer's cow. Or a legal quarry may escape in rugged or densely wooded terrain to die of its wounds days or weeks later. All these creatures are a bonanza for the fox, as is the offal left from dressing out large game before hauling it to the hunter's car.

Man's poultry and livestock also feed the fox, though more as carrion and debris than as food killed by the fox itself. There is no question but that foxes kill domestic fowl. A chicken or duck has no natural defenses against the fox, and birds which are left unattended and unpenned at night are an open invitation to a prowling fox. On the other hand, foxes hardly live up to their reputation of poultry-raiders—they kill far fewer domestic fowl than is commonly thought.

In Britain and other sheep-farming areas they are also accused of killing lambs, but the evidence that they do so is both scarce and unreliable. Sheep remains are found in fox stomachs, wool can often be discovered around a den, and foxes are sometimes seen feeding on lamb remains. But such evidence is chiefly an indication of the fox's efficiency as a scavenger. (Dead sheep play an especially important part in the fox diet in Australia, where one study showed sheep carrion to comprise 36 percent of the diet by volume over a two-year period.) During lambing time, especially in severe weather, a large proportion of lambs are still-born or die soon after birth. It is such dead or dying lambs which the fox takes, as well as the blood-stained wool clipped from the crutch of the ewe at lambing time, and the discarded afterbirths. There is no evidence of any significant loss of healthy lambs to foxes.

A striking way in which man has influenced fox living habits is the degree to which the animal has become a suburban dweller. Red foxes are found amazingly close to the center of New York City and they inhabit the outskirts of London. Through much of their range they are found at the edges of small towns. British and European suburbs are more congenial to foxes than many

American ones—in which an absence of hedges and well-planted gardens, and the presence of roads with high-speed traffic, cause hazards for a resident fox.

The fox has learned to be urban because urban man provides a lot of food for it. Garbage dumps are an obvious source, as are the rodents which inhabit them. Foxes have become nearly as adept as alley cats at raiding garbage cans—they have, in fact, been photographed doing so. Suburban parks and picnic spots provide scraps for the enterprising fox, as do highway truck stops and all-night cafés. The animals do not scorn bird food put out in backyards, just as they have no aversion to the birds for which it is intended. A nice suburban garden, an orchard, or a compost heap all have a great deal to offer the fox.

What is the overall effect of foxes in the ecology of an area? No comprehensive study has yet been made of this intricate question but certain points can be assumed. The fox feeds on such a variety of items that there is no constant and exclusive fox pressure on any one group of animals. Since the fox depends most heavily on food that is easy to get, it does not usually seriously affect food sources which are low or threatened with extinction. Foxes alone can have little effect on small-mammal populations when their numbers are high. Fox predation in the spring, however, when small-mammal populations are at a critical low, may play a role in preventing their rapid build-up. The fox has been shown not to have much effect on game birds, to be a scavenger rather than a killer of livestock, and to feed on poultry only occasionally.

Red fox population density is related to available food, of course, and also perhaps to the presence of suitable denning sites. Foxes have clear preferences for certain types of habitat, though small numbers can occur under a wide range of conditions. We know that the animals thrive and even increase conspicuously in agricultural areas. This is because farmland—mixed fields and woods—supports more rodents, birds, insects, and fruits than are found in heavy forest, especially established coniferous forest. A study in eastern New York State showed that the most common prey animals were associated with little-used or abandoned farmland, and none was of strictly woodland distribution. A Russian investigation which recorded the number of fox tracks in a variety of habitats found tracks most frequent at the edge of woodland and in young coniferous forests, and of moderate frequency in plowed land, deciduous brushwood, and non-arable fields and moorlands. Newly cut woodland and established fir forests yielded no track signs at all.

Figures are scarce on the numbers of foxes in a given area, and most figures, at that, are only educated guesses. Fox densities as high as 20 animals per square mile have been estimated for parts of Missouri. On the other hand, several years of direct observation in an area of Gloucestershire, England, found a ratio of about four adult foxes per square mile during the breeding season.

Over most of its range, the fox has no important predators except man and his dogs. Originally foxes and wolves occurred together throughout their respective ranges and the enthusiasm with which some domestic dogs hunt foxes today has led to the conjecture that wolves may in the past have had a decimating effect on fox numbers. But in the few places where wolves and red foxes now live in proximity, the two species seem usually to ignore each other. On Isle Royale, for instance, foxes benefit from wolf kills in winter and the presence of wolves does not seem to have harmed them. In Alaska's Mt. McKinley National Park, foxes and wolves in good numbers profit from each others' presence. Wolves enlarge fox dens for their own use, or visit the mouths of the dens to scavenge for scraps, while foxes feed from carcasses of wolf kills and raid wolf caches. While wolves do kill foxes if they get hold of them, they evidently do not hunt foxes systematically; foxes are far too small to fill a significant place in the wolf diet and far too difficult to catch.

Other predators that share the fox's range seem to have an equally insignificant effect on fox numbers. Coyotes will kill them but have few opportunities to do so. Bears relish meat but ignore foxes since they know they cannot catch them, and wolverines have little better chance. Wild cats, which in America include the puma, lynx, and bobcat, enjoy an occasional fox, though bobcats probably would not take on an adult; lynxes can catch foxes when deep snow impedes the dog but favors the large-footed cat. Big birds of prey occasionally account for a fox pup alone in the open. Eagles have been seen to swoop at adult foxes in Alaska; the fox takes up a defensive position, rigid and tense, with its tail straight up in the air, or if forced to run, eludes the eagle by dodging. Examination of birds' pellets in Alaska bear out the conclusion that foxes are rarely eaten by eagles.

It is man who is responsible for the deaths of enormous numbers of foxes each year. In the British Isles alone, some 50,000 foxes a year are killed by shooting, trapping, gassing of dens, and hunting with hounds. Great numbers are also killed in North America and the Soviet Union—more than 20,000 South Dakota foxes were shot from airplanes in the winter of 1971-72. In addition to hunting, man indirectly causes great numbers of fox deaths from highway accidents, live electric rails, and birds poisoned by dressed seeds. In eastern England during 1959-60, such birds killed some 1300 foxes within a few months.

By no means all of man's fox hunting is done to acquire the animal's valuable fur. Where hunting to hounds is a way of life, the fox is valued for the entertainment its pursuit provides; British foxes probably benefit from the cult of the fox hunter. In certain areas, of course, man sets out to destroy the fox because of its supposed role as a killer of livestock or game, or because it is a vector of rabies. Even when he wishes to wipe out the fox, however, man sets himself a difficult task. Despite the large number killed each year, the red fox survives and thrives throughout most of its range.

Like other members of the dog family, foxes are susceptible to a wide range of parasites and diseases. Most adults carry tapeworms, roundworms, and flukes, which to a point do not affect the animal's viability. Levels of parasitism seem to be influenced by the presence of sex hormones in the blood stream; parasites are at their lowest level during the breeding season. Tapeworm infestation appears to be related to the numbers of grass-eating animals the foxes consume; foxes which feed heavily on rabbits and deer carrion are more susceptible than those whose diet is predominantly mice. A large number of internal parasites are particularly harmful to young foxes and may be a significant source of juvenile mortality, while among adults infestation may simply aggravate weakness caused by a poor diet.

The fox has its share of external parasites—ticks, fleas, and mange mites. Fleas and ticks are more a nuisance than a mortal hazard, though heavy infestations can contribute to poor condition and spread internal parasites. Mange is another matter. In places such as Scandinavia it seems to occur rarely among foxes; in other areas, it can be so severe that through loss of hair it causes death from exposure during a harsh winter. The mange mite, which lays its eggs in the outer layer of the host's skin, can live in an abandoned fox den for up to two weeks. The disease is thus spread by animals entering infected holes. (Where foxes are commercially valuable, dens are sometimes treated with DDT and hexachlorane solution.) But the incidence of mange, like that of other infectious diseases, is lessened by the fact that foxes are not social animals and use dens for only a short time during the year.

Fox diseases include leptospirosis, distemper, canine encephalitis, and, most important as far as man is concerned, rabies. Rabies has little effect on foxes themselves, but the fact that they are carriers of the virus is of widespread concern to health authorities. Attempts in some parts of the U.S. to wipe out foxes, coyotes, skunks, bats, and all other rabies carriers have failed, however, because of the magnitude of the effort required. This is probably just as well, since the ecological problems brought about by the elimination of small rodent- and insect-eaters would probably be far more disastrous than occasional rabies epidemics, which are controllable in their danger to man by strict enforcement of vaccination measures on domestic dogs.

The social behavior of the red fox, quite different from that of jackals, coyotes, or wolves, is still far from being well understood. Some observers contend that adult foxes have little or nothing to do with each other except during the mating season, and that the male often plays no role in raising the young. Yet it is well documented that the male often not only associates with the female and pups, but aids in feeding the family. It is certain that adult foxes have not developed social mechanisms for group living, however, and usually keep to themselves much of the year.

The behavior of males and females differs during these solitary periods. The male fox is a surface dweller all year round; he is not bound to a den, but once he has established a home range he stays within a fairly well defined hunting territory. Scarcity of food may compel him to migrate to more favorable regions, and during the breeding season he is especially mobile in his effects to establish territory. This breeding territory is more restricted than the home range that demarks his feeding area, and is defended against other males during the rutting season. It is usually a mile or two in diameter, whereas the home range adhered to at other times of the year can be larger and may be shared by a number of other adults, all of which avoid conflict by scent and voice spacing.

The vixen tends to inhabit a more restricted range throughout the year and lives in her den during the breeding season. Vixens are sometimes found breeding quite close together, occasionally in the same den. Such neighbors do not form a pack; they seem to get along by a system of mutual tolerance rather than mutual aid.

Adult foxes establish social dominance and submission. In a given area one fox, often a male, is the dominant animal, with other adults ranged below it. Patterns of dominance change continually as new animals enter the area, old ones leave or die, or the mating season upsets the order previously established. Social structure varies regularly twice a year, once in the autumn, when animals are getting ready to breed, and again in the spring, when the requirements of that season wane and the males move about more freely.

Social rank is determined by various means, the most violent of which is actual fighting. Physical combat occurs most often between the dominant animal in an area and an intruder which has either inadvertently or purposely gotten in its way. The psychological advantage is on the side of the resident animal, with the intruder usually capitulating and fleeing before it is severely injured. Such fights are silent ones, accompanied by blood-letting and flying fur.

Combat may be more ritualized, with the contestants baring their teeth, snarling, erecting hairs on the neck, and engaging in a sort of pushing contest, paws on one another's shoulders. Or they may leap about each other in a kind of dance, sometimes bumping together before separating. This sort of contest is accompanied by open-mouthed screaming and yelling which can take place without any actual physical contact at all. When combat ends, the subordinate animal slinks away, tail between its legs. Conflict between males and females often seems to be restricted to screaming, the animals facing each other and vocalizing for as much as 15 minutes at a time. Pup play is also typified by fighting, leaping, pushing, and screaming—probably the way young foxes first sort themselves out in a social hierarchy.

The beginning of sexual activity in the fox varies widely according to latitude. Pro-estrus in the female may occur as early as November in such southerly parts of the range as southern Russia, and may take place in the far north as late as April. In the U.S. and Britian most matings are in January and February. Pro-estrus and estrus last from two to three weeks, during which the vulva swells and whitens somewhat but without noticeable bleeding. The female is receptive to the male and capable of being fertilized for from two to four days. Copulation is accompanied by the canine tie which commonly lasts from 15 to 25 minutes. The gestation period varies from 49 to 55 days.

We do not know whether or not the red fox is monogamous. Many experts think that in any given year the male usually mates with only one female, and remains with her during pregnancy and the raising of the young. In captivity, male foxes can be bred polygamously, but only if the male is allowed time to settle down with each successive female so that he comes to regard her enclosure as his own.

Roger Burrows believes that mating behavior is not a matter of monogamy or polygamy, but of territory. Evidence gathered during his study of foxes in Gloucestershire indicates that once the male establishes a breeding territory he will mate with females—whether one or several—within that territory. Where vixens' dens are close together the resident male may father several litters per year, whereas if den sites are more widely spaced he may stay monogamous. Males may mate with the same female during successive years but whether they do or not, Burrows feels, is an accident of territory rather than a matter of personal attachment. A male that loses its mate through accident or sickness will sometimes mate again in the same year if a female in breeding condition is available.

After mating, the vixen prepares her den for the arrival of the litter, either using the one she has favored during the breeding season or searching out a more suitable one. Dens are usually enlarged from holes originally dug by rabbits, badgers, or other animals, for, like many wild dogs, the fox is a good digger but finds it easier to improve an existing burrow than to start afresh. Dens are commonly found in loamy or sandy soil on well-drained, sloping hillsides. Because such places make good sites there are often several burrows close to one another, occasionally being used by several foxes at once. Dens are rather elaborate affairs with a system of tunnels and up to a dozen entrances, some of which may be excavations made by the pups. For no reason discernible to humans, vixens often change burrows after the litter is born or while the pups are growing.

The pups, numbering from one to as many as 12, but usually four or five, are born blind, with soft, thick, blackish-to-chocolate-brown fur. For the first month they stay underground, the vixen remaining with them for the first two or three days and then emerging periodically to hunt. As the weeks pass and the pressure of food-getting increases, the female spends longer intervals away from the den, and the pups are left to themselves for hours at a time. At about a month they begin to emerge from the den for short periods, first accompanying their mother to the entrance as she leaves for the hunt and then retreating, but later staying to play in the area around the den.

Once the pups are old enough to come out of the den, its entrance becomes littered with food scraps and droppings; the process of weaning involves the vixen's bringing the pups tidbits that she carries in her mouth, and also regurgitating food for them. When the pups are about six weeks old the vixen stops kennelling with them; she prefers to sleep by herself some distance away, although she continues to visit and bring food to them and the litter stays on at the den for many weeks more. The vixen undoubtedly removes herself in an effort to gain a little peace and quiet and to hasten the weaning process. (At the Tama Zoo in Tokyo, six-month-old pups have been seen suckling, which would certainly not occur in the wild.)

As mentioned, there is dispute over the role of the male fox in raising the pups. It was long taken for granted that before weaning the male helped by feeding the vixen and young. Captive male foxes have been seen giving food to females as well as carrying food into the den for the pups. Roger Burrows, however, thinks that though the male may occasionally associate with the vixen after the mating season, such a system is the exception rather than the rule. In his several years of keeping a close watch on fox dens, Burrows never saw more than a single adult—which in each case seems to be the vixen—enter the den to bring food to the pups. This negative evidence is not categorical proof that a male never approached the den or fed the pups, of course, but Burrows cites as corroborating evidence of his theory the fact that after the birth of the litter the vixen's weight drops conspicuously—indicating that she alone carries the burden of feeding both herself and her litter. Also, if Burrows is right about males mating polygamously within their breeding territory, a fox which fathers several litters in a season could not hope to be much help with each one.

Adolph Murie tells quite a different story as a result of observing foxes in Alaska's Mt. McKinley National Park. Like Burrows, Murie was well enough acquainted with the animals involved to be able to identify them as individuals, a prerequisite for interpreting social relationships. Murie's female, which he called Split-ear, had a litter of five pups and was attentively assisted by her mate. He hunted for the family and was eagerly and affectionately greeted by the female when he returned to the den. Though Split-ear's enthusiastic welcome might have had something to do with his bringing food, it seemed to Murie also partly pure affection.

Leonard Lee Rue also describes a close relationship between a denning pair of foxes in Alaska. Both parents brought food to the pups, but were away from the den so much that they encountered each other only every day or two. When one of the pair had brought in its kill and fed the pups it would sit down near the mouth of the den to look for its mate. If the other adult approached the den, the waiting fox would quiver in anticipation and then bound out to meet the incoming animal. Welcoming ceremonies seemed to be reciprocally ecstatic, with the foxes wailing, bounding about, waving their tails, and licking each other's faces.

In the case of Murie's fox family, a remarkable fact was the presence of several other adults. "There were at least three supplementary adults at Split-ear's den. . . . I saw from one to three on seven occasions. These nonparents, which functioned like maiden aunts or bachelor uncles, were treated passively by Split-ear and her mate—their coming and going meant little. When one made an appearance, it was noted and that was all. One evening, when Split-ear was impatiently awaiting the male, she met one of them and they stood together briefly and separated. The extra foxes brought food and called forth Split-ear's pups, played with them, and uncovered caches in the den area; they were right at home and behaved like parents. One evening, one of them baby sat while Split-ear hunted. It is not unlikely that these foxes were former offspring. Another possibility is that they had lost their own young and were satisfying their parental emotions. . . ."

Who is right, Burrows and others with their impressions of the red fox as a solitary animal which associates almost not at all with other adults except during mating? Or Murie and Rue, with their descriptions of affectionate mated pairs raising the young together, corroborated by radio-tracking evidence of close association of a male with his mate and pups during the early denning period? Both patterns may occur—as the result of different habitats, population densities, or pressures from man. Although experts agree that the case of the supplemental adult companions of Split-ear and her mate is unusual, the fact that vixens sometimes raise their pups close to each other or even in the same den indicates that different types of associations among adults are possible.

Fox pups develop fast. At birth they weigh from three to five ounces; male pups are generally slightly heavier than their sisters (species-wide, a few more males than females are born). Their eyes open after 10 to 15 days, and by the time they are four or five months old the pups are independent enough for the family to begin breaking up; sexual maturity occurs between 7 and 11 months of age.

The two most important activities in a young fox's life are eating and what people loosely call playing, an activity that is actually a combination of exploring the surroundings, learning to hunt, and scuffling and romping with litter-mates—to the accompaniment of much yapping and screaming. As the vixen

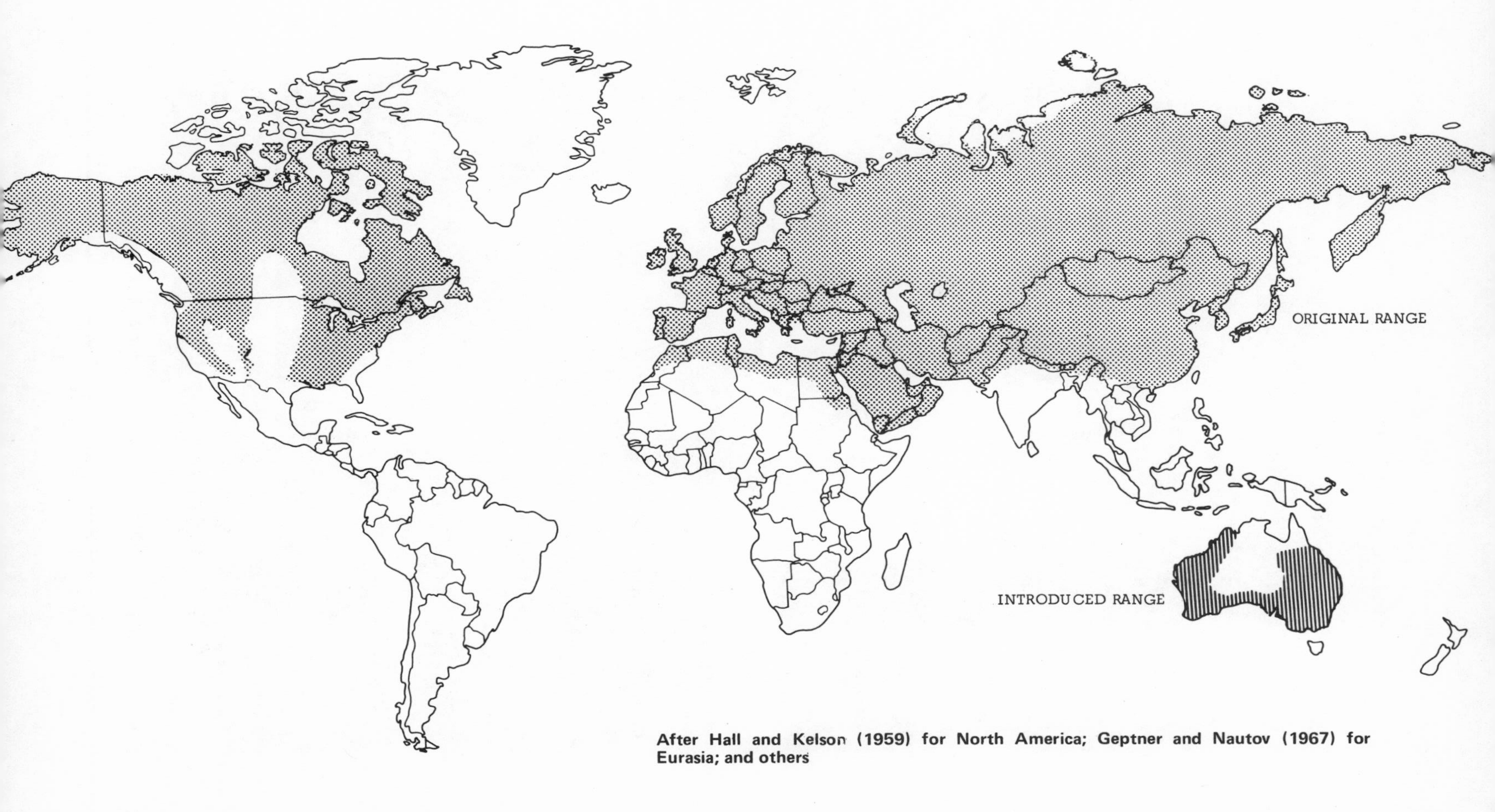

After Hall and Kelson (1959) for North America; Geptner and Nautov (1967) for Eurasia; and others

Red fox and pups at their den entrance. *Photo: Leonard Lee Rue III*

begins the weaning process the pup diet becomes increasingly varied. The adults even bring their young items they would themselves disdain, such as moles. But very quickly the pups begin picking up their own selections of small animals.

Fox pups are extremely curious, and this causes them to nose at, pounce on, and pick up anything which moves or has an interesting smell, as well as a lot of items which have nothing in particular to recommend them. If the object does not bite or pinch, and is not actively repulsive, it is mouthed, chewed, and swallowed–henceforth to serve as food if necessary. Pups eat great quantities of insects, earthworms, fruit, grass, and just plain trash.

As their strength, coordination, and experience grow they find larger prey such as rodents and birds, and the boldest of the pups are those which fare best in food collecting. There is no evidence that adult foxes teach the pups to hunt; they learn what they must through trial and error. During the summer and fall, after the pups have ceased to depend on the parents for food but long before they have reached adult size and sagacity, they are heavily dependent on types of food which can be obtained with little skill.

Fox puppy play, curiously enough, often includes the vixen. Among Burrows' foxes the favored playtime was dusk, and the vixen appeared regularly to lead the pups to a nearby playground–an orchard, a grassy field, or a hedgerow. After the play, which might last half an hour or more, she would lead them back to the den mouth before leaving on her nightly hunt.

As the pups grow, their play gets more and more vigorous, even violent. What starts out as a mock fight becomes a fight in earnest, and members of a litter sort themselves out in a hierarchy which is a precursor of the adult dominance pattern. The fact that male pups are usually larger and tougher than females affects the way the hierarchy forms. Sometimes there is a bond between the strongest and weakest members of a litter. Middle-ranking pups tend to behave more aggressively towards their inferiors than the top-ranking animals do, and the presence of the top pup in the litter seems to inhibit the grosser behavior of middle-ranking pups against the weakest. Thus the lowest-ranking animal may in effect receive protection from the dominant pup and seek out his company.

Young pups show their tendency for solitary living by playing or hunting by themselves, or catching naps away from their litter-mates. The break-up of the litter, which begins around the age of four months, is hastened by the increasing aggressiveness of play-fighting; in effect, the youngsters drive each other away as they become less tolerant of life in the group.

However winning fox pups may be, and however fascinating their parents, most people who go looking for foxes are in search of pelts. The prices of fox pelts rise and fall tremendously depending on whether long-haired furs are fashionable. In 1968 a good red pelt brought about $4.50 in the United States, a far cry from the very high prices of the 1920's but up from the pittances of the early 1950's, which made fox-trapping almost completely unprofitable. The real fortunes in fox furs were made by farms, which started breeding black and silver foxes in the 1880's and were the basis of a thriving business until the Depression and the market glut on furs drove most of them out of business by 1945.

The small trapper has continued his pursuit of foxes because even when fur prices hit bottom, he often has no source of income other than his traditional one. Fox trapping is done in the fall and winter; success is more likely either before the ground freezes or after the snowfall is heavy enough to hide a trap. Various sorts of sets are used. The blind set is put under a dirt or snow trail where the fox will step in it, or on a small rise onto which a suspicious fox will leap to scrutinize an unusual object set out by the trapper. But the dirt-hole set is the most common one. Foxes often cache their food, and other foxes commonly dig up such buried treasure. Knowing this, the trapper makes a small hole with something smelly in it that looks exactly like a fox's cache. In front of this hole, under loose dirt, he puts the trap.

Thanks to the trapper, to the effects of disease and accident, and to the rigors of surviving winters, fox mortality is very high. Tagged foxes in New York have lived an average of 178 days after being released; in Michigan they have survived for an average of 187 days. Most foxes in any population are not more than two or three years old. The fox needs its high fertility rate and its remarkably acute senses to survive. That it prospers over much of the world proves that it has that most important of canid traits—adaptability—to a rare degree.

THE KIT FOX (*Vulpes velox*)

In the American deserts and on the Great Plains lives the charming little animal called the kit fox. It would be more accurate to say that the Great Plains used to be the territory of the kit fox—for it has virtually disappeared from the central plains, though it still lives in the arid regions of the southwestern U.S.

In the early 1800's kit foxes were very common west of the line where forest gave way to the great prairies—from Alberta and Manitoba south through Texas into northern Mexico. Then about 1855 professional hunters started using strychnine to collect wolf and coyote pelts, and the kit fox was most frequently the victim of this indiscriminant poisoning. It was usually one of the first animals to feed at a poisoned carcass, and in the heyday of the poisoner, thousands were killed each year. Within 25 or 30 years the kit fox was almost eliminated from the plains, and the advance of civilization completed the process. Now the animal is generally confined to the most arid portions of the southwest and northern Mexico.

Two races of the kit fox have been distinguished—the northern or plains animal, the one now virtually extinct, has sometimes been called the swift fox to distinguish it from the desert animal. The northern variety has somewhat smaller ears and a darker coat than the southern and western animal, but otherwise they are very similar.

Like other desert foxes, the kit is small and looks delicate. Its head and body length is from 15 to 20 inches, with the tail another nine inches to a foot. It may stand as high as 11 or 12 inches at the shoulder but weighs only from four to six pounds. Its color is a pale gray, tan or sand, slightly darker on the back and with cream-colored throat, belly, and insides of the ears.

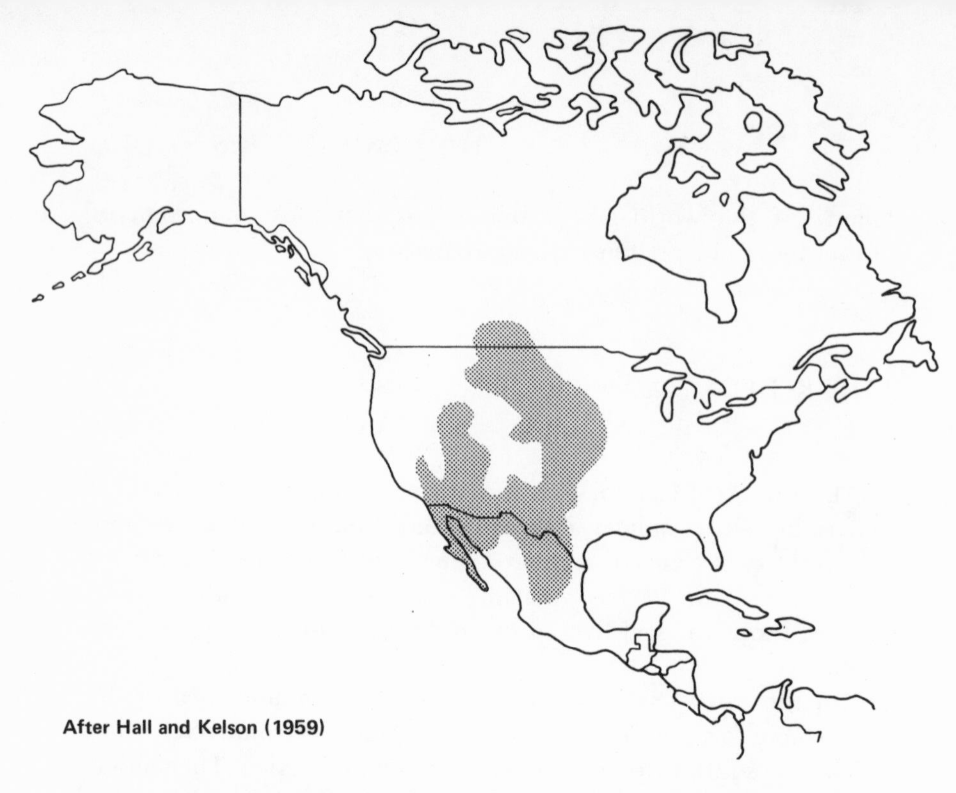

Kit fox, North America's desert fox. *Photo: Alfred M. Bailey, Denver Museum of Natural History*

There is usually a large brownish or blackish mark on each side of the muzzle and the tail tip is dark. The eyes are yellowish-brown, the large ears triangular and sharply pointed. Its voice—rarely heard—is a weak bark.

Kit foxes, which are nocturnal, live in permanent dens. Dens are often in rather barren places, frequently well-drained mounds slightly higher than the surroundings. In southwestern Utah, for instance, kit foxes usually den on ground sparsely covered with shadscale bushes, while in Texas they typically choose the most barren, over-grazed pastures. And where the kit fox population is high, a good denning area may have several dens clustered within an acre or two.

They often dig the burrows themselves, perhaps because the sites chosen have few rodent holes suitable for enlarging. New burrows frequently have only one entrance, but as years pass and families are raised, several more entrances and adjoining tunnels may be added. Old entrances plugged with dirt can often be found along with those currently in use. A typical feature of the den entrance is a ramp of dirt, up to a foot high, which extends several feet from the round opening. This ramp wears away with use and weather but is replenished with each renovation of the burrow. Also near the entrance, under a bush or rock, may be a slight hollow where a fox can lie during the day and still be within immediate reach of the den's safety.

Because the den is often in rather barren surroundings, the kit fox must leave its immediate vicinity to hunt the small animals that make up its diet. Within a mile or at most two will be an area, perhaps a fence row, sand dunes, or the edge of the foothills above a valley floor, where vegetation is thicker and rodents, insects, rabbits, and hares common. At dusk the foxes travel to these hunting grounds, returning to the protection of the den at dawn. Not unexpectedly they prey chiefly on nocturnal species like kangaroo rats and less heavily on such animals as chipmunks and antelope squirrels, which feed during the day. Like any other fox they are omnivorous opportunists, eating jack rabbits, rodents, reptiles, insects, ground-dwelling birds, eggs—and vegetable matter where it is available.

Although a good, steady supply of food must be within reach of this short-ranging fox, a permanent water supply is evidently not necessary. Many dens in Utah, for instance, are located miles from permanent water, so kit foxes must get enough moisture from their food. Animals in captivity drink regularly, however.

The breeding season for kit foxes in their present range is from December in Mexico through January and into February in Utah. Litters are born from February through April. There is no precise information on the gestation period, but it is presumably the 50 to 55 days normal to the red fox. From one to seven or eight pups may be born but the usual litter size is four or five. These foxes not only seem to be monogamous during the breeding season but appear to remain together the year around. The fact that the species uses a permanent den gives additional stability to the pair bond. For several weeks during and after the birth of the pups the burden of supplying the family with food rests on the male.

Pups grow rapidly and begin to appear at the den entrance when they are about a month old. At this age they are covered with short, woolly puppy fur, brown on the back and whitish underneath. On the shoulders and front legs the fur is becoming reddish and long black guard hairs protrude through the baby wool on the back and hips. Month-old pups have the short ears and muzzle of all young dogs, retain their babyish gray-blue eyes, and weigh a little over a pound.

For the next few weeks the pups appear outside the mouth of the burrow to engage in the usual puppy antics, scuffling with each other, digging frenetically in the sand or loose soil, catching and eating insects. They are not particularly wary of people, who have been able to walk right up to within 30 feet of a den, sit down, and have the pups come within a few yards. One student of kit fox behavior reports that a farmer plowing near a kit fox den for several days had found the pups paid no attention to him or to his tractor; when the machine actually crossed the den entrance, they merely took cover inside for a bit—and then popped out again when it had passed.

Kit fox pups stay at the family den longer than those of red foxes, and family groups frequently do not break up until autumn. Then the young disperse, some of them traveling significant distances. One young fox wandered 20 miles from the site where it had been tagged two months before. Sometimes a single fox den is the home of an extended family. One Texas den was occupied in July by five nearly grown pups, one adult female and two adult males. In other cases two lactating females have been found in the same den; perhaps they have been mother and daughter.

The kit fox is subject to the usual canine diseases and parasites and has its share of natural enemies. Coyotes kill them if they get hold of them and pups are sometimes victims of large birds of prey. Man is again the gravest danger, however. Some human destruction is inadvertent; in Utah, foxes like to sit on the warm pavement of highways at night and many are killed by cars. But also, since they are quite unwary of man, a casual rabbit hunter can approach within easy rifle shot of a den. And both traps and cyanide guns take a considerable toll.

When pressure from man decreases, however, kit fox numbers can rise rapidly. In Utah, new methods of coyote control involve wide spacing of poison stations and the placing of cyanide gun lines along accessible roads, which leaves large areas uncontaminated. As a result, kit foxes have increased in the past 20 years and in some places can even be called abundant. One hopes that more areas will be made safe for the species, since a more pleasant and innocuous little animal can hardly exist.

THE CORSAC FOX (*Vulpes corsac*)

The steppes and deserts of Central Asia are the home of the corsac fox, also known as the steppe or tartar fox. The corsac is found primarily in the Soviet Union, from as far west as the northern edge of the Sea of Azov to as far east as the Transbaikal steppes. It reaches north to Omsk in western Siberia, while the southern edge of the range may enter northern Iran. It extends into northern Afghanistan, extreme northwest China in the Dzungaria basin of Sinkiang, Mongolia, and the northern edge of Inner Mongolia in northeast China. Within the Soviet Union the corsac fox is in many places the most common predator.

Smaller and less foxy smelling than the red fox, the corsac varies from 19 to 24 inches long over head and body, with the tail an added nine to 14 inches. Some animals weigh as little as five lbs., though heavier individuals are found. Like other desert foxes the corsac has relatively long, slender legs; the ears also are quite long, very broad at the base, and with such a sharp point that they form a distinct triangle. The skull, as is to be expected from the overall dimensions of the animal, is noticeably smaller and lighter than that of the red fox.

The color of the corsac is basically a reddish-gray or reddish-brown, with a silvery shade to the back because of the light-tipped guard hairs. Color is more intense along the middle of the back and shoulders, and becomes lighter on the flanks, with the belly a yellowish or off-white. The backs of the ears are the reddish-gray of the back itself, and the tip of the tail is dark brown or black, not the white of the red fox.

The corsac prefers a rolling, open, arid habitat, and shuns heavy forest, thickets, and populated areas. In years when fox numbers are high, however, it may be found in the forest steppe at the fringes of its range. It lives in burrows, usually those dug by badgers, marmots, or red foxes, though when necessary it will dig a simple shallow den for itself. Where corsac foxes are numerous they form "corsac towns"—clusters of burrows quite close together.

Active chiefly at night, the corsac is shy and cautious. This behavior may be due to intensive hunting in the wild, for in captivity, corsacs are especially popular with zoo visitors: they quickly become tame, and are lively even in the daytime. In 18th-century Russia, in fact, they were kept as pets. The corsac is extremely agile, has an astonishing jumping ability for its size, and also climbs well. The corsac's agility is of little use in deep snow, however, where its slender legs and small paws cause it to flounder and sink.

The corsac feeds chiefly on rats and mice. In summer, it also hunts birds and eggs, and will catch such small creatures as jerboas, ground squirrels, hedgehogs, hares, and insects. It sometimes hunts other small predators like mink and polecats. Adult marmots are too big to tackle, but young marmots are common prey. Because of its own small size, the corsac requires a relatively small amount of food. Animals in captivity have been found to eat an average of nine or ten ounces a day and rarely to eat more than a pound.

From year to year there may be great variations in the numbers of this species, depending on the availability of voles and mice. In places where prey is common or in years when diseases attack voles or marmots, the fertility of the corsac fox increases markedly and foxes come from other regions. In such times of plenty the fox may store food; during one year of vole decline in Mongolia, a fox burrow was found to contain 80 voles and 11 other small mammals. But when prey is scarce, the corsac fox weakens, its fertility decreases, disease spreads, it becomes an easy target for other predators, and its numbers drop abruptly.

Little is known of the family life of the corsac fox in the wild, though we do have information from zoos. Estrus occurs in January and February; fighting goes on among the males at this time. Although the gestation period is usually reported as between 50 and 60 days, at the East Berlin Zoo, gestation periods have ranged from 49 to 51 days, which corresponds with the 51 or 52 days usual for the red fox.

Litter numbers range from two to 11 and probably average from four to six. There is on record one litter of 16 young, though it is possible that in such a case two females might have littered in the same den. It is also possible that, since the corsac fox resembles the arctic fox in its dependence on fluctuating populations of small rodents, it might be similar to the latter in having a widely variable fertility, and be capable of producing very large litters in good years.

The grouping of corsac fox dens into "corsac towns" indicates the sociability of this small predator. The male fox seems to have an active role in raising the litter. (When the East Berlin Zoo left the males of their corsac fox group with the females during and after whelping, the father also took an active part in defending the young against presumed enemies.) In the wild, young foxes stay with their mother at least until fall and, according to some information, through their first winter.

The corsac fox is widely hunted in the Soviet Union for its fur, though it does not have as thick a pelt as the red fox. It is taken by trapping, shooting, and digging up burrows. The number of corsac foxes in the Soviet Union has declined in recent years both because of natural causes such as disease and as

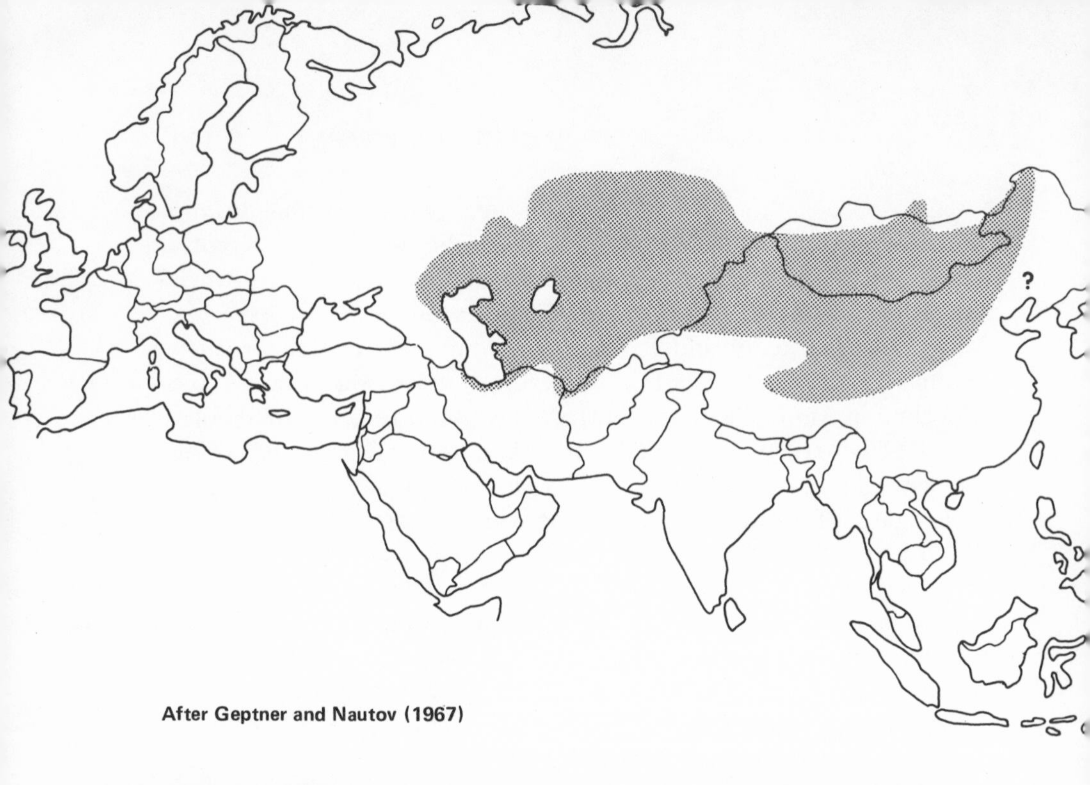

After Geptner and Nautov (1967)

Corsac fox with pups. Corsac foxes are native to the steppes of central Asia.
Photo: Gerhard Budidi, Berlin Zoo

a result of heavy hunting, but this animal is still among the most important furbearers in its range.

THE BENGAL FOX (*Vulpes bengalensis*)

In addition to various races of the red fox, the Indian subcontinent has a species all its own: the Bengal or Indian fox. The Bengal fox is found from the foothills of the Himalayas (up to about 4500 feet) to the southern tip of India, and from the Indus River in Pakistan on the west to eastern Bengal and perhaps through Bangladesh into Assam on the east. The skull and skeleton closely resemble that of a small red fox, but externally the animal has several features that clearly set it apart. The tail tip is black, the backs of the ears are colored like the back itself and are never black, and there are two or three pairs of mammae rather than the four pairs common to the red fox.

The Bengal fox is relatively small—17 to 23 inches for length of head and body, and ten to 14 inches for the tail. Males weigh from six to eight pounds. The color is a speckled gray on the back, with the top of the shoulders and the crown of the head brownish, reddish, or buff. The muzzle is dark and there is usually a grayish-black smudge in front of the eyes. The under parts are whitish with the exception of a dark chin, often a dark band on the lower throat, and a reddish area around the anus and genitals. The backs of the thighs and outsides of the forelegs are reddish down to the hock. The tail is dark grayish and ends in a black tip.

As is true of foxes in general, the color of the Bengal is richest when it is in full winter coat, and fades during molt and in the summer. Though it develops a woolly undercoat in the parts of its range where the winters are cool, it nowhere has very heavy fur—the hair of the soles of its feet, especially, are thin and short.

The Bengal fox is an animal of open areas and scrub thickets. Since it is not obnoxious to man and goes unmolested by him, it is comparatively fearless and often wanders around villages and dwellings. It is an active, swift-footed animal which, though by no means able to outrun hounds, can often elude them by its

astonishingly quick dodges. The voice is a sharp, chattering bark repeated three or four times. This fox probably gives off less of an odor, at least from its foot glands, than the red fox, since British hunters in India have reported that hounds easily lose its trail.

Unlike some of its kin, the Bengal usually digs its own den, choosing relatively high ground in areas where flooding occurs. The burrow can be a fairly complicated affair, with several entrances and passageways leading to a central chamber.

Rodents, ground birds, eggs, lizards, land-crabs, and insects, especially termites, form the diet of the Bengal fox. It has been seen catching grasshoppers and beetles and even jumping into the air for moths. It also eats fruits like melons and the plums of the ber. It is reportedly not much of a pest to poultry-keepers; perhaps wild prey provides a sufficiently steady diet.

Bengal foxes mate from November to January, depending on the latitude and elevation, and the gestation period is presumed to be 50 to 55 days. The litter frequently numbers four. The female stays close to the den while she is suckling the pups, which are reported to live at the burrow until they are nearly grown.

THE HOARY FOX (*Vulpes cana*)

The hoary fox is an almost unstudied small animal from central Asia. Also called the Baluchistan fox, Blanford's fox, or Bukharian fox, it is found in the southern part of the Turkmen S. S. R., in northeastern Iran, through Afghanistan, and into the Baluchistan and North-West Frontier Provinces of Pakistan.

Very small, the hoary fox has a tail which may be almost as long as its body. The length of head and body is under 20 inches (usually closer to 16), while the tail ranges from 11 inches to as much as 16. The ears are also long for the size of the animal—two and one-half to three and one-half inches. Especially in winter, the hoary fox has a soft, densely furred coat with a very full tail. Its color is light gray on the upper side, with a darker brownish or blackish tinge along the back, while the undersides are whitish. The chin and lower lips are dark and the backs of the ears are the same dark gray as the back. The legs may have some black but usually are uniformly buff or ochre. The tail is gray with black tips to the hairs, and the tip of the tail itself is black.

Nothing whatever seems to be known about the habits of the hoary fox. G. A. Novikov, who examined several specimens from Turkmenia, notes that the species is one of the rarest predators in the Soviet Union. R. I. Pocock was able to study several prepared skins, but measured and examined only one specimen in the flesh. It may well be that the animal is not as rare as it seems, at least in the center of its range. However, since Afghanistan and adjacent areas have not been thoroughly studied by biologists this small fox has pretty much escaped attention.

THE TIBETAN SAND FOX (*Vulpes ferrilata*)

The high plateaus of Tibet are the home of the Tibetan sand fox. Knowledge of the animal is sketchy and imprecise; its range certainly falls mainly within Tibet but it occurs in northern Nepal and perhaps also in the upper Sutlej Valley of extreme northwestern India. Within Tibet its habitat is evidently the high,

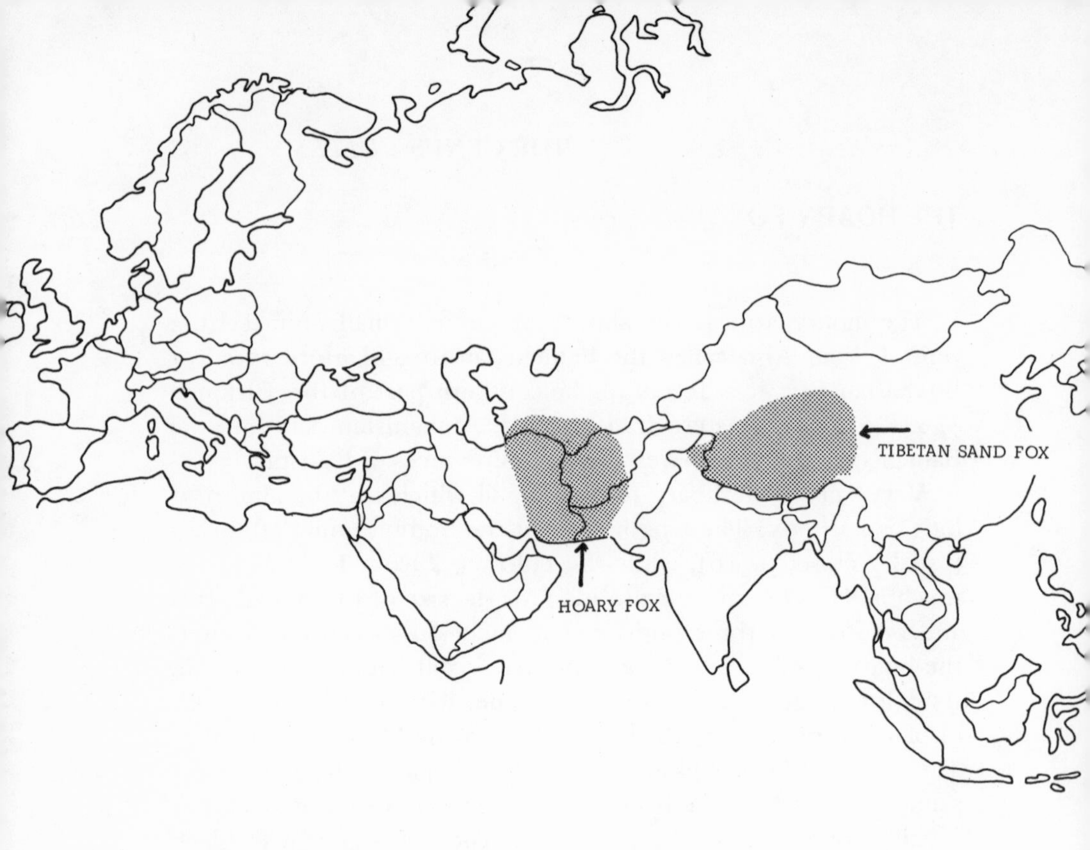

treeless plains above 12,000 feet. Its biology is completely unstudied, though we may assume that it lives mainly on rodents and other small mammals. For instance, it has been seen catching rockhares (black-nosed pikas).

The Tibetan sand fox is like the red fox, and unlike other Asian foxes, in having a white tip to its tail. Its size, and relatively short ears and tail, make it quite unlike the Bengal and hoary foxes. The mounted specimen described by R. I. Pocock had the following dimensions: head and body 26½ inches, tail 11¼ inches, ear two and one-fifth inches.

Like many animals which live at high elevations, this fox is notable for the very great thickness of the fur which, however, is rather short, the guard hairs being generally no longer than the underlying wool. The back is a uniform reddish or yellowish; interspersed with the fur are some thin black hairs, but they are not enough to change the overall light color. The head is often somewhat grayer than the back. The backs of the ears may be a little darker than the nape of the neck but they are not black. The sides of the neck, the flanks and the thighs are silvery, the belly and chin are white. There is some reddish or buff color around the hock on the hind leg and the elbow on the foreleg but

Tibetan sand fox, captured at about 12,000 feet elevation near Namche Bazar, Nepal. *Photo: R. Marlin Perkins, photographer; copyright Field Enterprises Educational Corporation*

the paws are white. The dark gray tail has an extensive white tip.

In addition to the color and external characteristics such as the relatively short tail and ears, certain cranial peculiarities of the Tibetan sand fox are distinctive. Compared with the north Indian race of the red fox, which it resembles in overall size, the Tibetan fox has a somewhat longer skull. The difference lies in its relatively longer, narrower jaws. The teeth are also different, the canines being longer and the molars of different proportions. From what little we know of it, the Tibetan sand fox seems to be a distinct species found in a distinctive habitat; it is clearly an animal about which it would be interesting to know a great deal more.

RÜPPELL'S SAND FOX (*Vulpes rüppelli*)

A typical desert-dwelling fox, Rüppell's is an attractive, small, sandy-colored creature with the enormous ears so often found among mammals in hot climates. It is an animal of North Africa and the Near East. In North Africa it is found from Morocco to Egypt, occurs in the Ahaggar, Aïr, and Tibesti ranges of the central Sahara, and reaches into the northern Sudan, Somalia and northeastern Ethiopia. It occurs on the Sinai peninsula and is found in the Near East from as far north as Iraq south throughout the Arabian peninsula and east to Afghanistan.

Although its size, color, and ears have caused it to be confused with the fennec (see page 182) Rüppell's fox shows anatomical and genetic characteristics that link it closely with the red fox. Its head and body length ranges from 17 to 19 inches; the tail is disproportionately long—about 11 to 12 inches. It stands about ten inches high at the shoulder and weighs some six lbs. The very large ears are three and one-half to four inches long. Allowing for its smaller size, more delicate build, and bigger ears, the skull and teeth of Rüppell's fox are similar to those of the red fox.

It has a soft, fine coat with very dense underwool. Hair so covers the soles of the feet that it almost completely conceals the foot pads, an adaptation of obvious use in a hot sandy habitat. The color of the coat is the pale neutral tone one expects in a desert animal. The back, with the exception of a contrasting orange dorsal stripe, is a paler or darker gray which varies according to the individual or the subspecies; for instance, the race found on the Arabian peninsula is conspicuously paler than the animals from North Africa. The back and upper flanks are lightly speckled with black hair tips and the black speckling becomes a little heavier on the tail, which otherwise continues the buffy or grayish tone of the back. A light but distinct reddish-orange stripe extends down the back from the neck to the base of the tail, and continues somewhat paler on the tail itself, which ends in a conspicuous white tip. The belly and the insides of the legs range from buff to almost pure white. The face is distinctively marked by reddish-orange spectacles around the eyes and black patches in front of them.

These coat markings, as well as features of the skull, readily distinguish Rüppell's sand fox from the fennec. The whitish tail tip of the former is another difference, for the fennec sports a black tip.

Work on the chromosome numbers and characteristics of some members of the dog family has helped to clarify the relationship of Rüppell's fox to the other vulpine foxes and to

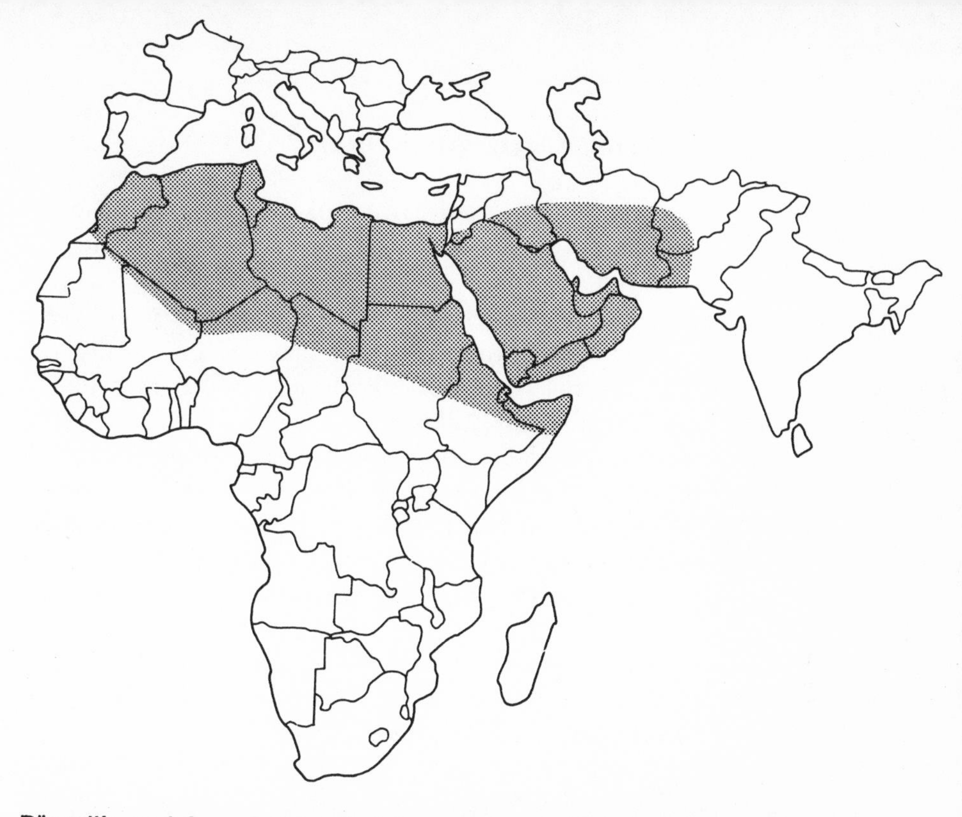

Rüppell's sand fox, photographed in southwestern Egypt. Facial markings, speckling on the back, and the white tail tip distinguish it from the fennec.
Photo: Xavier Misonne, Belgian Scientific Expedition to the Libyan Desert

the fennec. Rüppell's fox has 40 chromosomes, more than the red fox but similar to them in appearance. The fennec, on the other hand, has 64 chromosomes. Thus, although Rüppell's fox and the fennec look so much alike externally that they have been confused constantly in the literature, they are genetically and anatomically distinct and present an interesting example of parallel adaptation to desert life.

Little detail is available about the habits of Rüppell's fox. It seems to be restricted to the deserts and steppes within its range, and presumably lives primarily on rodents and perhaps insects. It has been encountered in northern Saudi Arabia in places where rodent signs are common, or where garbage has been dumped.

THE AFRICAN SAND FOX (*Vulpes pallida*)

The African savannas south of the Sahara are the home of yet another obscure little fox called, simply enough, the African sand fox. Similar to other vulpine foxes in desert habitats, its range extends east to the Nile River in the Sudan, west to the interior of Senegal. In the south the sand fox goes to the point at which savanna changes to tropical vegetation; to the north it is found along the edges of the desert and occurs in the Marra, Tibesti, and Aïr mountains of the Sahara.

The African sand fox has thin, short, pale sandy fur. There is sometimes black speckling on the midline of the back and on the tail, which usually sports a sharply contrasting black tip. Two adult males in captivity weighed three and three-quarters and four lbs. respectively. The ears are less exaggerated than those of the other north African foxes, Rüppell's and the fennec.

Although the behavior of this fox is little known, three young animals from Chad, a female and two males, were acquired by the Nürnburg Zoo in 1965, thus allowing us a little information. They were fed a diet of mice, mealworms, and biscuits and adjusted quickly to both climate and captivity. In June of that first year, the female gave birth to a litter of four, three of which survived. All three adults were together in the same enclosure when the young were born, and seemed to get along amicably.

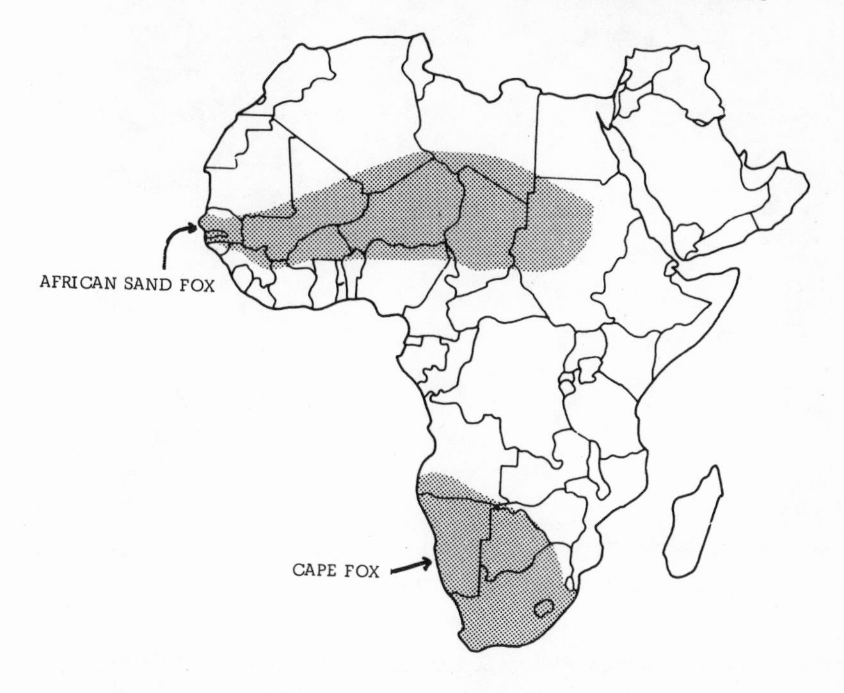

The rather feeble male assumed to be the bachelor kept apart, but the presumed father joined the female and pups in the box during the first day.

On the day after birth one of the pups, a female, weighed one and two-thirds ounces. The young foxes developed rapidly and at five weeks old ranged in weight from five to nine ounces. At 14 weeks of age the smaller male weighed two lbs. 10 ounces, the larger two lbs. 15 ounces, and the surviving female two lbs. 5½ ounces. The breeding success of the Nürnberg Zoo leads to the hope that the African sand fox will eventually be bred and studied at other zoos.

THE CAPE FOX (*Vulpes chama*)

The Cape fox is the southern African representative of the vulpine foxes. It has a variety of common names: Kama fox, ass fox because of its long ears, silver fox from its color, and silver jackal as a translation of the Afrikaans name *silwerjakkal.* (The

A pair of African sand foxes from Chad. *Photo: Nürnberg Zoo*

Cape fox of southern Africa. *Photo: San Diego Zoo, photo by Ron Garrison*

Cape fox is not, of course, a true jackal, but in Afrikaans the word *jakkal* is applied to most wild dogs.)

The Cape fox is found in the western and central regions of southern Africa, from southern Angola and southwestern Rhodesia south across Botswana to northern and western Transvaal, Orange Free State, northern Cape Province, and parts of South West Africa. The presence of man has confined it to the more arid portions of this range.

A small animal measuring about 24 inches for head and body, the Cape fox has a thick, bushy tail of 13 inches or so. It stands about 12 inches high at the shoulder and weighs eight to ten lbs. Its ears are long for its size, some three and one-half inches, and the muzzle is small and sharp. The color along the back is silvery as a result of the mixture of colors on individual hairs; the base of each hair is a dull gray which shades to a much lighter band in the middle, with the tips of the long guard hairs black. Where guard hairs are thin or absent, as on the neck or sides, the color seems lighter than on the back, since more of the light middle band of each hair is exposed. The head of the Cape fox is a dull red-orange, with a brownish tone to the lower jaw and angle of the mouth. The backs of the ears are tawny and the long hairs on the insides whitish, with a dull yellowish marking at the front base. On the back of the hind leg, between the knee and the heel, there is a clearly defined black band. The hairs of the brush-tail are basically yellowish tipped with black; the tail tip itself is dark.

A shy, cautious animal, the Cape fox is primarily nocturnal and appears to begin its hunting at dusk. If caught away from the den by daybreak it shelters under bushes, rocks, or in ditches. It prefers open plains and veld and is seldom found in wooded regions. Its voice is a sharp bark like that of the red fox. While the red is conspicuous for the heavy, musky scent it leaves, an odor noticeable even to the insensitive nose of man, the Cape fox characteristically leaves little scent.

If disturbed or hunted the Cape fox makes for shelter as fast as possible. At full speed it stretches out low, seemingly with its belly almost touching the ground. When chased by dogs it is capable of remarkably quick, sharp turns, apparently aided by the use of its long tail, and can frequently tire out a pack of hounds and eventually evade them. It inhabits burrows which, rather than dig for itself, it borrows from springhares, aardvarks, or meer-cats. The Cape fox has the variety of internal and external parasites usual among dogs, and seems especially susceptible to distemper.

Small mammals such as rodents, rock hyraxes, and hares are fair game for the Cape fox, as are smaller creatures like ants, locusts, beetles and their larvae, and ground-nesting birds and eggs. It will catch lizards, eat wild fruit and berries, and dig out juicy roots and tubers. Also conspicuous as a scavenger, it is commonly seen searching among the debris of deserted campsites. Such a habit makes it easy to trap and leads it to pick up poisoned bait set out for other animals.

Like other foxes, the Cape fox does not shun stray poultry and will occasionally raid a farmyard for fowl. In parts of South Africa it has been hunted and poisoned by farmers who think it is a danger to sheep—an unfounded charge, though it does feed on sheep carrion.

The gestation period is 51 or 52 days, with the young born in August and September, early spring in southern Africa. The average litter size is three to five pups. Both parents cooperate in raising the litter and carry rodents and birds to the weanling pups, also burying small animals near the den as a cache of ready food. The presence of pups in a burrow is often betrayed by the pieces of skin, feathers, small bones and hares' feet scattered around the opening.

Arctic Fox

(Alopex lagopus)

THE ARCTIC FOX lives exclusively in the northern polar regions. Together with the polar bear it ranges out onto the pack ice of the Arctic ocean, farther north than any other land mammal. Other wild dogs, for example the timber wolf and the red fox, range quite far north, but they are typically animals of the temperate zone. Only the arctic fox, sometimes called the blue or white fox, is exclusively an arctic canid, and it alone of all the wild dogs experiences a seasonal color change.

Its range is circumpolar. Basically its habitat is the arctic tundra, but in addition to roaming north onto the icefields it also moves south into the forest tundra in some areas, especially during winter, and inhabits all the major arctic islands. On the continental land masses the southern-most point of its range varies; it goes farther south along the edges of the ocean than it does inland. In the Soviet Union the fox generally ranges north of 67° to 70° latitude, though it extends farther south along some treeless river valleys and open ridges, and on the coast of the Kamchatka peninsula. It is native to the relatively southern Commander Islands and has been successfully acclimatized on some of the Kurile Islands. In western North America it is found throughout the tundra, down into the forest tundra, and on the Aleutian and Pribiloff Islands. In eastern Canada it ranges quite far south along the coast of Hudson's Bay and has been found at the mouth of the St. Lawrence River.

The arctic fox reflects its range in its physical characteristics—luxuriant fur, long and very bushy tail, small ears, short legs and face, a color change to winter white which minimizes heat loss. All these are attributes of an animal particularly well adapted to severe cold. Body length varies from about 19 inches to as much as 31; with the tail, an additional ten to 12 inches. The animal stands 12 inches or a little less at the shoulder, and weighs from six to 15 lbs., with 12 or 13 lbs. the normal winter weight.

Compared to the red fox, the arctic one seems set low to the ground, its head is relatively rounded, and the rounded tips of the ears protrude only slightly from the winter coat. The soles of the feet are completely covered with thick, coarse hair. The coat itself is wonderfully dense and soft; in winter, besides being at its thickest, it develops air cavities within the hairs which add to its insulating properties.

The color of the arctic fox is particularly interesting. In addition to experiencing a seasonal color change, the animal has two distinct color phases in the winter coat. The *winter white phase* is by far the more common, although it appears to be genetically recessive. In this phase the animal is completely white during the winter, and in summer is dark brownish or grayish on the back and yellowish-white on sides and belly. In the *blue phase* the winter coat is dark, ranging from light gray through slate blue to nearly black, or sometimes from yellowish to dark brown. Some bluish or grayish animals may have brown on the feet and head. The summer variation of the blue phase is like that of the white phase, though occasional blue animals retain their winter color the year around.

The ratio of white-phase animals to blue-phase ones varies greatly from place to place, and the reasons for the variations are still not well understood. Two facts are clear: one, that blue foxes are most common at the fringes of the arctic fox's range, such as the Pribiloff Islands and other southern island groups, and two, that foxes living in areas where lemmings are abundant are usually white while those living in areas without lemmings are more often blue. These two observations correspond with one another since foxes in non-lemming areas usually live along the seacoast, where they feed primarily on sea birds and cast-up debris, and many of these seacoast areas are on the islands and continental edges of the fox's range. The question remains,

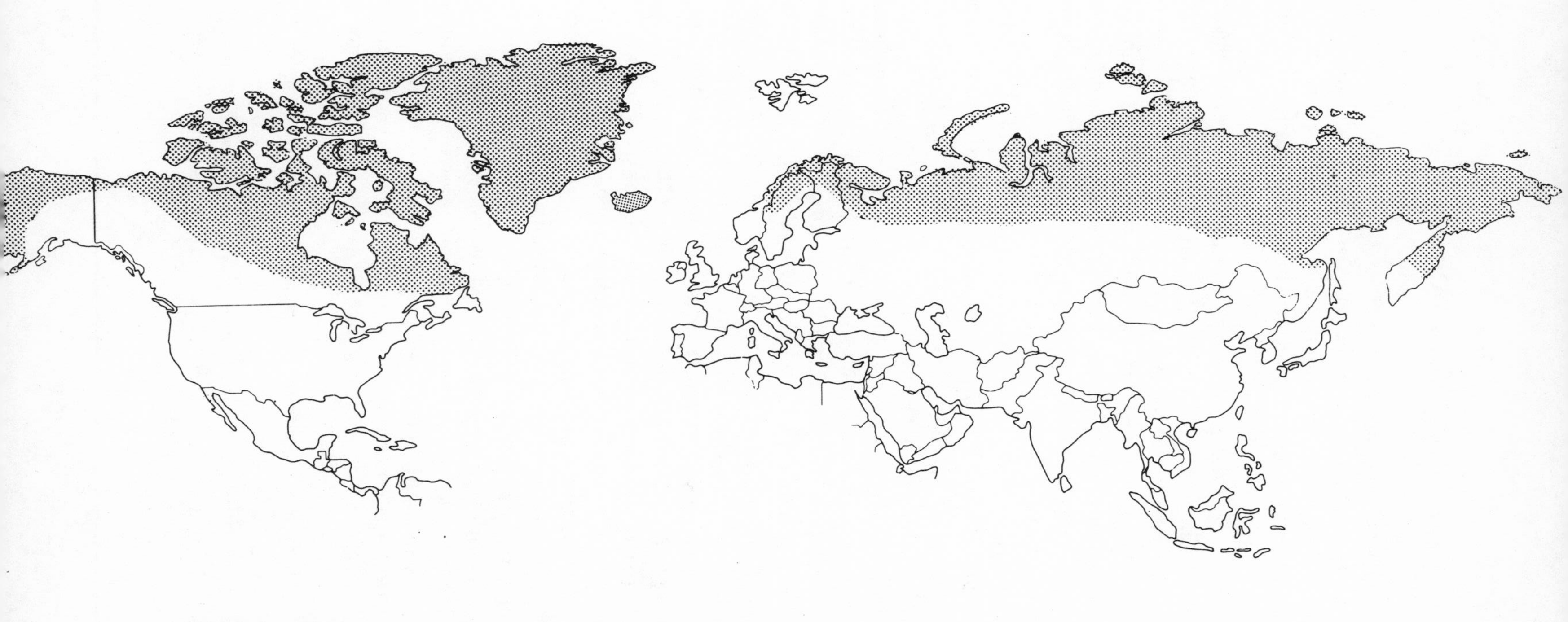

After Hall and Kelson (1959) for North America; Geptner and Nautov (1967) for Eurasia; Hildebrand (1954); and others

however, whether diet itself affects the color, or whether the more severe winters of the interior of large land masses, where lemmings happen to be found, favor the white phase while the blue phase is better adapted to the milder climate of the sea coast.

In any event, the proportion of blue foxes is very high, about 90 percent, in such places as Jan Mayen Island and the vicinity of Thule, Greenland, where the major source of food is the small auk called the dovekie. On the Pribiloffs and the Commander Islands the proportion is also very high, and has been increased by the trapping activities of man, who has turned both areas into a sort of natural fur farm where foxes are trapped live and the white animals removed so that the more valuable blue phase is further increased. But in the interior of Siberia, only three or four percent of the arctic fox population is of the blue phase and in the Canadian arctic even less.

The seasonal color change in the arctic fox is accomplished by a twice-yearly molt. In mid-spring the hair begins to thin, the long guard hairs on the flanks falling first, then the undercoat beginning to come out in clumps. Simultaneously the brownish or yellowish summer coat, considerably thinner than the winter one, starts to grow. In fall the fox molts again and new hair grows in white—summer hair does not simply turn. Winter white begins to appear first on the belly, then on the flanks, and finally on the back, hastened or delayed by the fox's range and by the weather. The young whiten less quickly than mature animals.

The arctic fox is neither nocturnal nor diurnal, moving about at all hours both in summer and during the darkness of polar winter. But its movements are affected by the weather, and during storms or in unusually cold or windy weather it stays in the shelter of its temporary den or snow burrow, sometimes for days at a time. As behooves a coastal and island-dweller, this fox is a good swimmer. Its voice resembles that of a small dog, sometimes a soft bark or growl, at other times a rather harsh bark.

The fox moves agilely at a walk or a trot. Lois Crisler describes the movements of two animals frolicking. "They were the lightest things on their feet we ever saw. A cat is light but he makes preparations for his effort—crouches, then springs. The foxes made no preparations. They did not bounce, they did not spring. They 'floated.' Touch, touch, touch, like soap bubbles." The delicacy of its movements is enhanced by the fact that the animal's broad, furred paws and light weight prevent it from sinking into the snow.

An interesting arctic fox characteristic is its boldness in the presence of man. An early group of Arctic explorers who were wrecked on Bering Island in 1741 were pestered nearly to distraction by hoards of inquisitive and thieving foxes. They followed people about, entered their dwellings, stole equipment and clothing, and ravished food supplies. So eager were they to steal food that they were often accidentally stabbed while the party was skinning seals. In the ensuing 200 years arctic foxes have learned to be somewhat more wary of man, but accounts of animals which come close enough to be touched or to feed from the hand are still common.

The most important sources of food for the arctic fox are rodents, chiefly lemmings and field voles, and (for those foxes which live along the coast) sea birds, seashore debris, and dead whales, seals, and walruses. The fox is a scavenger that follows both man and the polar bear, and feeds also on the remains of wolf and wolverine kills. In summer it raids birds' nests for eggs and young, and eats berries too. In the Maritime Territory of the Soviet Union, along the coast of Greenland, and on the arctic islands, birds and shore animals such as sea urchins, mollusks, crustaceans, and fish remains compose the bulk of the fox diet.

On the continental land masses where rodents are all-important to the fox, its fecundity and survival vary with the availability of lemmings and voles. In a good rodent year the number of fox stomachs empty of all food may be only four to five percent; in a bad year this figure can rise to 24 percent. Mortality during the first year of the arctic fox's life is as a rule very high; when rodents are scarce almost no pups survive. When the lemmings and voles again increase, however, females bear and raise large litters and the fox rebuilds its numbers.

Arctic fox hunting methods vary according to where they live. Those which depend heavily on sea birds must make the bulk of their catches in summer, when the birds nest along cliffs. They usually kill far more than they can eat at once and cache the remainder to tide them over the winter. Such caches are not inviolable, of course: foxes eat from them even in summer; nor are they usually sufficient to provide all the food during the months when the sea birds have departed, and in late winter and spring emaciated foxes can be seen waiting patiently beside pools of open water, where high tides may break up the icebergs, hoping to obtain a tidbit. By March, when the polar bear leaves its den to begin its hunting, the fox is eager to follow the larger predator and feed from its catches.

Among the lemming-hunting foxes of the continental land masses, there is less difference between the plenty of summer and the scarcity of winter; here the abundance of food depends on the yearly fluctuations of the lemming population. Even in winter foxes can catch lemmings, which build their neat little houses under the snow but above the frozen ground. The foxes hear or smell the rodent beneath the snow, then leap straight up in the air to dive down headfirst through the snow crust and onto the lemming house below. When the snow melts in May the lemmings are even more vulnerable, for the ground is still too hard for them to escape by digging. This season of spring plenty coincides for the fox with the raising of its young and when lemmings are abundant the period of the melt-off is an easy one for the little predator.

Among arctic foxes estrus occurs earlier or later in the spring, depending on latitude, the weather, and the physical condition of the animals. In the more southern parts of the range it takes place in February or March, with signs of estrus appearing among Commander Islands foxes in late January. On Southhampton Island in northern Hudson's Bay, mating takes place in March and April, while in some parts of Siberia it may be as late as May. Arctic foxes lack the bloody discharge which marks pro-estrus in the members of the genus *Canis,* but the period preceding mating is characterized by an increase in social activity among the animals, including much playing between the sexes and a three-to-five-tone bark from alternate individuals which evidently serves to bring animals together.

Early in the breeding period the female is followed by several males which sometimes fight quite savagely among themselves. Once two animals pair off they are apparently monogamous for the season and the male remains with the female through the birth and raising of the litter. The gestation period for the arctic fox can vary from 49 to as many as 60 days, with the average at 51 or 52 days.

The fecundity of the arctic fox depends on the kind of diet and the abundance of prey animals. Coastal foxes produce relatively small litters, averaging about six pups and rarely exceeding ten. The foxes of the interior which prey on lemmings respond to years of rodent abundance by producing very large litters—sometimes as many as 20.

Though pups are occasionally born among bushes or clumps of driftwood, arctic foxes typically bear their young in dens, which the females often inhabit even before the breeding season. The most desirable burrow sites are either in heavily forested tundra among lakes, or along the raised shores of rivers, lakes, or the sea. In such areas the ground is neither too moist and swampy nor too solidly frozen to allow digging, and food is likely to be plentiful. By preference foxes build their dens several miles from those of their neighbors, though where sites are scarce they may be much closer. Burrows are re-used every year, renewed and broadened by each new owner.

At birth pups weigh two or three ounces and are covered with short velvety brown fur. They begin to open their eyes at 14 to 16 days, and emerge from the den by the time they are three weeks old. Their diet consists exclusively of milk for the first month, supplemented by meat later on, and the pups are weaned at six to eight weeks. The male assists in feeding the pups as soon as they can receive solid food, and he probably feeds the female as well.

By the time they are ten to 12 weeks old the pups accompany their parents in hunting easily-caught prey—rodents and molting or young birds. The parents stay with the pups as they begin to learn to hunt; first the male and then the female leaves the litter, which usually stays together in the home den until five or six months old. By fall the litter has broken up, and the young foxes are on their own. Arctic foxes reach puberty early, at nine or ten months of age, and mate at the end of their first year. Once they survive the crucial first year, when the mortality is from 60 to 80 percent, they have a life expectancy of eight to ten years.

It is possible that there is some contact and communal rearing among female foxes in areas where burrows are scarce or have to be close together. In a group of foxes observed in captivity two females shared a den while one was raising a litter and the other was pregnant. The female with the litter was tolerant of the maternal behavior of the pregnant female, which shared the maternal tasks of her neighbor by following and retrieving the pups and lying around them in the nursing posture. When the second female gave birth, both mothers cooperated in retrieving each other's pups.

In external appearance and habits the arctic fox differs considerably from a fox of the genus *Vulpes.* The skulls and skeletons of arctic and red foxes, however, are very similar, as are genetic characteristics. Fur breeders have crossed the arctic and the red fox, hoping to combine arctic fox fertility with red fox size and color phases. Hybrids can be produced but they are completely sterile, thus frustrating the desire of breeders to backcross. The arctic fox has 52 chromosomes, while the red has 34. The chromosomes of the hybrid are unable to pair properly to produce viable sperm or egg, so sterility results. But the fact that the red fox and the arctic fox can be bred and regularly produce offspring indicates a closer relationship than the classification in separate genera suggests.

In overall economic importance the arctic fox is one of the most valuable furbearers, and is shot and trapped throughout its range. In some places the more valuable blue fox is raised on fur farms. The Commander Islands of Russia, and the Pribiloffs of the United States, are, in effect, fur farms where the animals live wild but are given supplementary food and are selectively live-trapped. The arctic fox appears to be in less danger than many other heavily hunted animals, perhaps because man, however eagerly he may pursue the animal, has no desire to inhabit its range in any significant numbers. So long as its habitat remains unspoiled it appears able to tolerate considerable hunting pressure.

The arctic fox must also be wary of animal predators—timber

Arctic fox in winter-white coat. Its short ears and muzzle help it conserve body heat. *Photo: Leonard Lee Rue III*

wolves, wolverines, and other species of foxes—which share its range. A polar bear will try to catch a fox if it ventures too close or is cornered. But more destructive than predators is disease, which attacks arctic foxes as they become abundant in response to an increase in the lemming population. Rabies and encephalitis are the most common diseases, but worm infestations also become a menace as the numbers of foxes grow. Migration, a reaction to population pressure, takes another toll as foxes move to other areas in search of food.

The arctic fox is an excellent example of a predator whose fortunes rise and fall with those of the rodent population. It shows very clearly the typical predator adaptations to such a situation—high fecundity in good years and small litters or widespread barrenness in lean years, high pre-natal mortality and a high death rate during the first year, rapid maturation, and prevalence of diseases during periods of heavy population pressure. Such a pattern results in a flexible species able to take advantage of or adapt to changes in its environment.

Fennec

(Fennecus zerda)

THE FENNEC is to the desert what the arctic fox is to the polar regions, an animal conspicuously well adapted to its extreme environment. It is sometimes even known just as "the desert fox." It is found across North Africa as far south as the Aïr Mountains of Niger, and Kordofan and Sennaar in southern Sudan. Its range crosses the Red Sea into Sinai and the Arabian Peninsula but it is not as common there as has previously been thought, almost all references to the fennec in Arabia being the result of confusion with Rüppell's sand fox (see page 166). Though it looks very much like desert-dwelling vulpine foxes, the fennec is accorded separate generic status because of its several specialized features.

It is distinctive, first of all, for its small size. From nose tip to tail base, it measures from 14 to 16 inches, with the tail an additional six to as much as 12 inches long. It generally stands eight inches at the shoulder and weighs about three lbs. Its ears are quite remarkably large, four to six inches long and very wide at the base. It is these enormous ears—at least half the height of the animal's body and, relatively, the largest among dogs—which are its most conspicuous feature.

A fennec's skull has several characteristics which distinguish it from the smaller vulpine foxes. It is conspicuously smaller than that of Rüppell's fox, which shares its range, yet in comparison with the latter the braincase is striking large, smooth, and

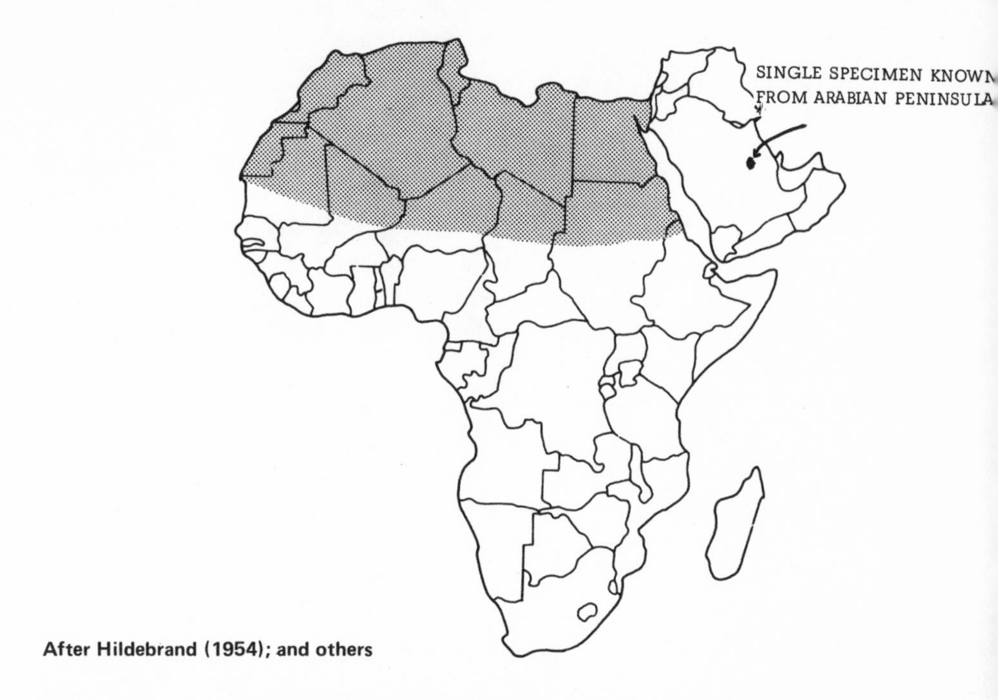

After Hildebrand (1954); and others

Fennecs, pale, tiny North African animals with huge ears and pointed faces, are perfectly adapted to desert living. *Photo: San Diego Zoo*

inflated. The tympanic bullae (the bony prominences below the ear holes) are very large; the muzzle, on the other hand, is delicate. The teeth are light and rather feebly developed; the canines, especially, are relatively small and slender.

Fennecs are very pale in color, some animals being almost white on the back, others fawn, and some a reddish buff. The underparts are a very pale buffy white. The tip of the tail is black, as is the nosepad, and there is a black spot on the upper surface of the tail near the base. In contrast to the light fur the eyes seem extra dark and prominent.

The fennec has soft, fine, thick fur, the longer guard hairs covering a dense, woolly undercoat which serves as insulation against temperature extremes. The edges of the ears are heavily fringed with long white hairs, and the soles of the feet are so densely covered that the pads are invisible. The fluffy coat, small size, pale coloring, and winsome face all combine to make the fennec look like a cuddly toy.

However cute the animal's appearance may be, it is the result of severely practical adaptation to a rigorous environment. The fennec must withstand extraordinarily high daytime temperatures in its desert home, and tolerate rapid and extreme changes of temperature between day and night. The prominent extremities, such as the pointed nose and the huge ears, effectively radiate excess body heat. The light color serves as partial camouflage in desert sand, but it also absorbs less heat from the overpowering sun. Not even the fennec's natural adaptations allow it to brave the mid-day desert sun. It is a dawn-dusk and nocturnal animal, not just because it must avoid man, as is the case with so many so-called "nocturnal" predators, but of natural necessity.

Active and agile, the fennec is fast over short distances. A captive animal has been known to jump two feet straight up from a standing position and to make a horizontal leap of four feet. The fennec's voice, heard in captivity, is a bark like that of a small domestic dog. Little is known about its life in the wild, but since it is an appealing animal and easy to tame, the fennec has frequently been kept in captivity, especially by Europeans living in North Africa. One man kept a pair quite successfully in his Paris apartment, fed them a diet of meat, eggs, fresh fruit, and cookies, and found them tame, if annoyingly hyperactive at night. They were easy to housebreak to a box of dirt, which would lead one to believe that their habits are equally fastidious in the wild.

The fennec uses a permanent den which acts as shelter from enemies and climate as well as a place to rear the young. It is fully capable of providing this den for itself, for it digs so rapidly that it is said to look as though it were sinking into the sand. The fennec is far more sociable than many foxes. Sometimes several dens are found close together, or even interconnected, and captive fennecs appear to enjoy each other's company. But although they sometimes live in extended family groups, fennecs hunt singly.

Like other foxes, the fennec is truly omnivorous, and its small size allows it to subsist very well off the small animals which are typical of deserts. Insects are particularly important in the diet, migratory locusts being a favored and occasionally plentiful item. Small rodents like gerbils are eaten, along with birds, birds' eggs, lizards, tubers, and other vegetable matter. Fennecs are said to enjoy dates and to be able to climb date palms to reach the fruit. It is easy enough to imagine their scrambling along a bent or leaning trunk, though almost certainly they do not scale very tall, straight palm trees. They drink freely when water is available and sometimes numbers gather at a waterhole at night, but it is possible that they are able to go without water for extended periods. Fox tracks have been reported in the desert far from oases.

Fennec litters are born in March or April after a gestation period commonly estimated at 50 or 51 days, though the Strasburg Zoo has had two litters after observed gestation periods of 62 and 63 days. The young usually number from two to five. Judging from the communal living habits of the species, young fennecs may stay with the parents for a long time and then perhaps eventually make their own dens nearby. Though little is known of fennec biology, the animal has been found to have 64 chromosomes, many more than the 34 to 40 found in *Vulpes.* This fact sets the fennec apart genetically as well as in size, proportions, and cranial characteristics from the other desert foxes.

Gray Fox

(Urocyon cinereoargenteus)

THE MAJOR INDIGENOUS FOX of much of the U.S., the gray fox is a handsome, distinctive small animal found from southern Canada to the northern part of South America. Called the "chacolillo" in Central America, it occurs as far south as Venezuela and recently was recorded in Colombia as well. There are several subspecies of the gray fox, some of which have been given specific status in the past, but the genus is now generally considered to have only one, albeit far-ranging and varied, species.

A little smaller on the average than the red fox, the gray gives a rather different field impression. It is more upstanding, with proportionally longer legs and shorter muzzle, ears, and tail. The length of head and body is from 21 to 30 inches, with the tail an additional 11 to 16 inches. The gray fox stands 14 to 15 inches at the shoulder and weighs from seven to 13 lbs., with an average of eight to ten. An inhabitant of relatively warm regions, this fox has rather coarse fur and lacks a dense woolly undercoat.

Despite its name, the gray fox's coloring is often bright. Its upper side, from nose to tail, is a salt-and-pepper gray, rich and dark among more northern animals, somewhat washed out farther south. The sides of the legs, underside of the tail, sides of the neck and back, and base of the ears are a reddish-orange which is sometimes quite vivid and contrasts handsomely with

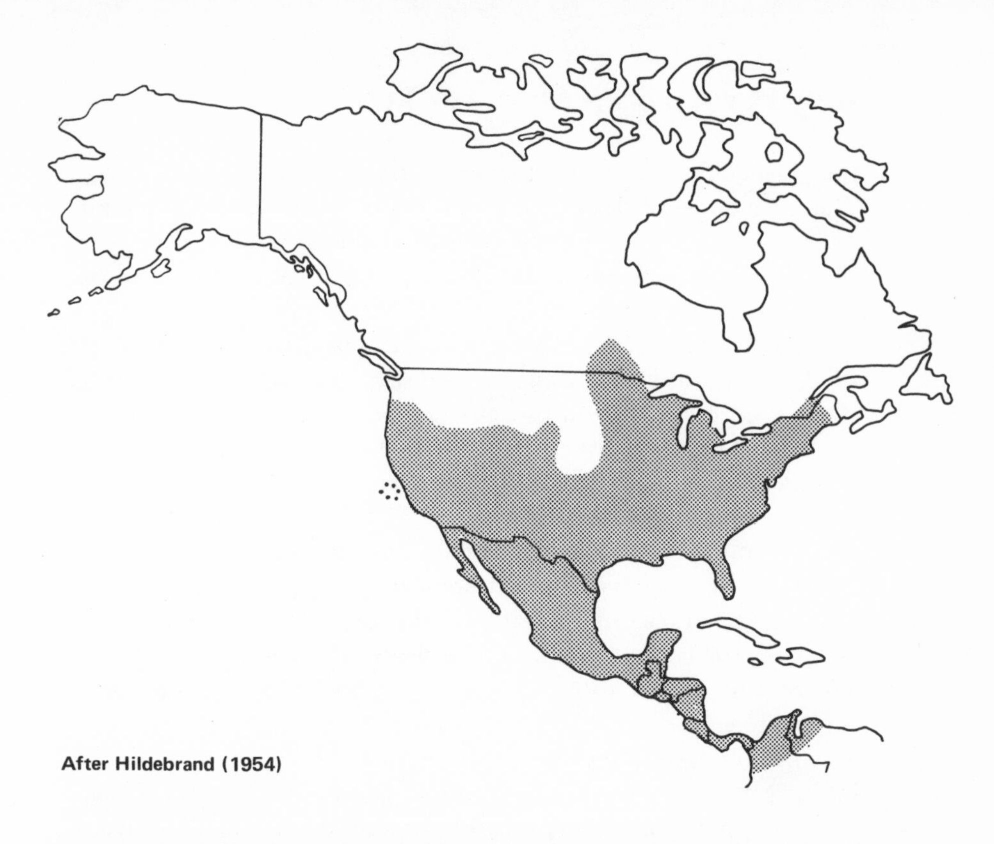

After Hildebrand (1954)

The North American gray fox climbs trees to hunt and to escape enemies.
Photo: Copyright Paul W. Nesbit 1968

the gray back and the buff or white belly and insides of the legs. The nose and the sides of the muzzle are black, and a black line extends from the outside corner of the eyes back to the neck. The tip of the tail may be blackish or merely an extension of the back's gray; it is never white.

This fox has certain distinctive anatomical features. The pupil of the eye is oval, not a narrow slit as in the vulpine foxes, and there is an erectile band of hairs on the upper side of the tail. Its skull can be distinguished from that of the red by several rather technical characteristics, and it appears also to lack the strong musky smell of the red fox.

The animal has its own forms of fox-like behavior. The voice is a harsh yap or bark which sometimes sounds like that of a coyote. The tracks of the gray are similar to those of other foxes as far as gait is concerned, but its proportionally larger toe pads and overall smaller foot size distinguish its tracks from those of the red fox. In snow it seldom shows a tail-drag.

When disturbed or hunted by man or dogs, the gray fox is unlikely to take off on a long cross-country run; instead it holes up in a den or disappears into heavy trees or brush. It is aided in its evasion techniques by being able to climb trees and is the only wild dog which regularly does so. It scrambles up tree trunks by gripping with its forepaws and forcing itself higher with the long claws on its hind feet. The descent is a backing-down process. The gray fox is sufficiently at home in trees to be able to leap from branch to branch when hunting birds or other prey, and to take to the trees and hide among the foliage when pursued.

Before the extensive forest-cutting and farming of the white man changed the face of North America, the gray fox was found along the Atlantic coast as far north as New England, across the forested central and southern part of what is now the United States, and in the partially wooded portions of the southwest. In much of this range it was the only fox, as the red fox did not originally occur much south of New England and the kit fox is an animal of the treeless plains and desert. With the extension of farming in the north Atlantic states the gray fox became less common in the northeast and came to be thought of as an animal of the more southern pine woods, where it has always flourished.

In the past few decades, however, the abandonment of

marginal farms in the northeast and the reversion of such land to woods has encouraged the return of the gray fox. At the same time it has expanded northward in the Rocky Mountains and along the Pacific Coast. The gray fox has never favored a completely treeless habitat and thus is not found on the Great Plains or the upland deserts of the intermountain west. It is well adapted to the sparsely timbered, arid uplands of much of Central America, though it is also found to sea level and in both heavier forest and farming land.

The gray fox inhabits a permanent den which may be a hollow log or a tree, a rock cave or a crack. It may den in a hole in sandy ground, but such a den is borrowed, not dug by the animal itself. It is sometimes found under buildings in secluded places. The den is frequently lined with shredded bark or leaves, and may be used year after year—either by the same fox or a series of different animals.

Most active after sundown, the gray fox returns to its den during the day. It feeds on rats and mice, hares and rabbits, squirrels, eggs and birds, all sorts of insects, and a variety of fruit: persimmons, wild cherries, wild grapes, and berries. Like other small dogs it eats reptiles and amphibians when it finds them. Because it climbs, arboreal creatures like squirrels and young birds and eggs are more important to it than to most wild dogs. It may also ambush passing prey from overhanging branches along a trail.

Gray foxes mate at the end of their first year, with the mating season falling sometime in the spring, according to latitude. In the central and southern United States, mating is usually completed between late January and the middle of March, though in New York State it may take place as late as May. The gestation period averages 63 days. There can be from two to eight pups in a litter; three to five is average.

Newborn pups are black, weigh about three ounces, and have their eyes tightly closed. The male stays with his mate and helps support her and the pups, which are weaned at about six weeks. The young remain with the parents for several months, gradually learning to fend for themselves. They first leave the immediate den area to hunt with the mother when they are about three months old. They drift away in late summer or early fall, though the parents may stay together until the new mating season approaches. Gray foxes have been reported associating in small packs, but these are probably only immediate family groups. (This animal, however, is considerably more social than its solitary cousin the red fox.)

Annual mortality among gray foxes is very high. A study of tooth wear in Georgia gray foxes showed that about 60 percent were under a year old, 28 percent were yearlings, seven percent were two-year-olds, and only a little over three percent were three or more years old. The gray fox suffers from many of the same diseases and parasites as the red fox, though its more social habits and the fact that it lives in a den the year round make it more susceptible to epidemic diseases. For instance, gray foxes are far more subject to encephalitis than are reds; this is a viral disease easily spread by contact with infected animals and contaminated dens.

The gray fox is less famous for its adaptability to the proximity of man than is the red, though the fact that it is extending its range in the United States, and perhaps in South America, indicates that it is not seriously threatened. It is an easy animal to trap, but its coarse, thinnish fur does not have much commercial value. By climbing it can easily elude stray dogs and unskillful hunters and as a lover of arid regions it will probably continue to prosper throughout much of North America.

The Genus Dusicyon

There is less known about the wild dogs of South America than of any other continent, and describing them can be a confusing business. The gray fox is found in Venezuela and Colombia but is primary a North American animal; other than that, the wild dogs of South America fall into several groups. Placed in separate genera stand the bush dog, the maned wolf, and the small-eared dog. That leaves a large group of medium-sized to small dogs, all rather similar, which occur almost everywhere on the continent. These animals, popularly called South American foxes or jackals, seem to combine, both in appearance and behavior, some of the traits of those two dogs of the northern hemisphere.

South American foxes are referred to under a list of bewilderingly various common and scientific names. There has been only one encompassing study of any consequence, that of Angel Cabrera and José Yepes in 1940, along with a revision of classifications published by Cabrera in 1957. Cabrera's taxonomy has been generally followed by subsequent writers and is the basis for most of the following information. Though Cabrera's work is more complete than any other, it leaves a multitude of questions unanswered—and raises several new ones. One hopes it will not be long before the wild dogs of South America get more detailed and systematic attention.

Cabrera divides the South American foxes into two genera, most of them included under *Dusicyon* and one sufficiently distinct to be called *Cerdocyon.* The species included in the genus *Dusicyon* are found in the temperate south of the continent and up the Andean mountain chain as far as Colombia. They are slender foxes with large ears; the tail is long and well furred, with a black spot on the upper surface near the base of the tail, which itself always ends in a black tip. The legs are slender, with spreading toes and long, narrow nails. The forehead is flat, and the angle of the lower jaw very narrow. Within the genus there are two groups, the gray and the red foxes or, as they are known there, the "true" foxes and the culpeos. Although Cabrera distinguishes eight separate species, I have chosen to group the representatives of *Dusicyon* into four sections, each containing one clearly distinctive species and one additional similar species which subsequent study may or may not accord specific status.

In addition to living representatives of the genus *Dusicyon* one more species, from the Falkland Islands, has been described. This animal, *Dusicyon australis,* variously called the Falkland Island dog, Falkland fox, Falkland wolf and Antarctic wolf, no longer exists. Although it was exterminated by man nearly 100 years ago, it is mentioned in many accounts of South American fauna. When the first settlers arrived in the Falklands they found the fox to be the only land mammal. It fed on birds, and perhaps also on seals and other shore animals. The settlers had brought sheep with them, and naturally launched an all-out attack against the predator. An easy target, it had become extinct by 1875.

The question has always been how the Falkland fox found its way to these islands 400 miles east of the Straits of Magellan. One possible answer is that the islands are within the continental shelf and may once have been connected with South America by land. It would have been strange if only the fox had used the connection, but perhaps it subsequently eliminated any smaller mammals which once lived there. It is also possible that the fox reached the Falklands on drifting ice during the glacial age.

THE CULPEO (*Dusicyon culpaeus*)

The several species of the genus *Dusicyon* fall into two fairly distinct groups, the gray and the red foxes. The culpeo, as it is known in Chile, is the chief representative of the latter group. Called the Andean wolf in Ecuador, the culpeo is found in the Andes from Colombia to the southern tip of Chile, eastward in Argentina throughout the Patagonian plateau, and south to Tierra del Fuego. It is an animal of the temperate zone; while in the southern part of the continent it ranges to sea level, farther north it is restricted to the mountainous regions, high deserts, plateaus, and upland savannas. The subspecific differences among culpeos have not been thoroughly studied, though the chief variations are in size and color. The Tierra del Fuego race seems to be the largest and there may be a steady south-to-north diminution in size.

Even so, the largest member of the genus, the culpeo, is only a medium-sized animal—about the same size as the African black-backed jackal. Body length varies from 29 to 33 inches, with the bushy tail an additional 15 inches, though the Tierra del Fuegan race can be as much as 60 inches from nose to tip of tail.

The coat of the culpeo is a mixture of yellowish and black, with black predominating along the back. The tail, somewhat more yellow than the back, ends in a black tip. The jaw is grayish-white, becoming darker gray towards the end. (This feature helps to distinguish the animal from those members of the genus known as the gray foxes, all of which have blackish jaws.) The name "red fox" comes from the rusty markings on forehead and jaw and the even brighter red found on the backs of the ears and up the outsides of the legs. The coat is not thick enough to be of great commercial value; the pelts of animals from southern Patagonia have the longest, softest hair and find a readier market than others.

The culpeo's habits and diet are similar to those of other medium-sized wild dogs. It is rarely seen during the daytime, though it is evidently less shy and wary than smaller members of the genus. Where food is scarce it may range widely at night. It

Culpeo, the so-called "red fox" of South America. *Photo: San Diego Zoo*

depends on small mammals, birds, and lizards for food, preying chiefly on the Andean rabbit in Ecuador and in Patagonia taking birds—even such large ones as the bustard. It is widely reported to be a killer of sheep, especially at lambing time. Other details about the culpeo are lacking, though its name, a Chilean word meaning "madness" or "folly," was given to it because of its reputation as an easy target for hunters.

The Santa Elena Fox (*Dusicyon culpaeolus*)

Were it not for its location, the Santa Elena fox would undoubtedly be considered a member of the preceding species. This animal, isolated in a very small area of southeastern Uruguay, is separated from the culpeo of the southern Andes by several hundred miles of Argentine pampas. Though somewhat smaller, it has the reddish facial markings and dirty white jaw that characterize the culpeo and distinguish it from the "gray fox" members of the genus. The anomolous location of the Santa Elena fox is explained by the fact that during the late glacial age, the Argentine pampas were inhabited by foxes of the same type as the present-day culpeo which, while they disappeared from most of their pampas range, have hung on in the departments of Soriano and Rio Negro in southeastern Uruguay.

THE PAMPAS FOX (*Dusicyon gymnocercus*)

Chief representative of the South American "gray foxes," the pampas fox is also called the common or Paraguayan fox. Compared with some other South American wild dogs, it is relatively well known, because its range is the pampas around the large population centers of Argentina, Uruguay, and Brazil. The southern race of the pampas fox ranges the pampas of Argentina from the western provinces of Cordoba and San Luis to the coast, and south as far as the Rio Negro; the northern race, which is more vividly colored, occurs in Paraguay, northern Uruguay, and southeastern Brazil from the state of Paraná to that of Rio

Pampas fox of the South American plains. This fox is distinguished from the culpeo by its dark muzzle. *Photo: Francisco Erize—Bruce Coleman, Inc.*

Grande do Sul. The pampas fox is called "guarachaim" by the Brazilians and "aguarachaí" by the Guaraní Indians of Paraguay and northern Argentina.

A medium-sized animal, its body measures 31 or 32 inches, with a tail 13 or 14 inches long. The fur of the gray coat is long enough to make the pelt acceptable in the manufacture of cheap fur garments. The back is mostly black mixed with a pale yellow, while the tail has more yellow on it and carries the two black spots typical of the genus, one on the upper side at the base, and the other at the tip. The throat and belly are whitish and the jaw black as far back as the corners of the mouth. The muzzle, ears, and neck at the base of the ears are a rusty red, as are the outsides of the legs. On the backs of the hind legs the rust of the lower leg is separated from the brindled thigh by a black band.

The pampas fox is well known for two behavioral peculiarities. One is its collecting habit—it picks up and stores or hides seemingly useless objects, especially bits of cloth and leather. In the days before it was hunted to its present scarcity, travelers on the plains, or peasants and gauchos who slept in the open air, had to take precautions at night to see that the foxes did not steal pieces of bridles, lassoes, and other tack. On the pampas, such objects are sometimes found in burrows of the rodent known as the viscacha; they may have been put there by the foxes which so often take over the burrows as dens.

When approached by a human, the pampas fox "freezes" or plays dead. If it is encountered unexpectedly and believes itself unnoticed it will throw itself to the ground rigid, with eyes closed. The famous traveler and naturalist W. H. Hudson tells of having seen a fox in such a position actually beaten with a rawhide whip, yet remain motionless until it thought itself abandoned. Possibly in such death-feigning the animal does not completely lose consciousness but is so terror-stricken that it is insensible to punishment. Exactly what happens physically when the fox plays dead is not known, nor is it certain why the animal does so, though the behavior is certainly involuntary. Though agile and a good dodger, the pampas fox is not, however, very fast on a straight-away and cannot hope to outrun a horse, say, so perhaps has adapted this method of saving itself when cornered.

The voice of the pampas fox is often heard after dark. The cry is a single sharp bark; if it is heard in quick succession from several different points, a number of foxes are probably contacting each other; such a chorus of cries is more common during the mating season than at other times of the year. Although primarily nocturnal the pampas fox may range abroad during the day in solitary places where it does not expect to encounter man.

As its name implies, the pampas fox is a plains animal. By preference—even where there are forests within its range, as in Brazil—it seeks out the open country. It prefers to den in cover of tall grass, dense low undergrowth, or a field of crops. It regularly inhabits a den, which it prefers to borrow from an armadillo or viscacha rather than dig itself. Apparently the pampas fox sometimes simply moves in and takes over an occupied viscacha burrow, forcing the rodents to leave. Viscachas are large, sturdy animals the size of big hares, and usually resist such invasions, if rather unsuccessfully. If the intruding fox is a female which raises a litter in the burrow, she may, with the help of the young, eventually wipe out the whole viscacha colony.

Pampas foxes like to eat rodents of all kinds, especially field rats, and also prey on birds like partridges and small herons. When food is scarce they hunt frogs and lizards. They have a definite taste for fruit and other vegetable matter, especially the lower parts of sugar cane stalks. Near populated areas they are reputed to be great despoilers of poultry yards.

Basically a solitary animal, the pampas fox is usually seen in pairs only during the mating season, though occasionally two are found together during the summer. Foxes mate in the southern hemisphere's winter, from July to early September, and pups are born in October or early November. The length of the gestation period is evidently unknown. A litter usually consists of three to five pups, which are born almost black and gradually lighten as they mature. The pups spend the first several weeks at the den but by the time they are from two to three months old can accompany the parents on hunting trips.

Very little is known about the social behavior of this fox. A study of them in captivity at the London Zoo indicates that, despite the name, they act more like the European jackal than like foxes of the genus *Vulpes.* For instance, the pampas fox holds its tail out stiffly when expressing dominance, as do the members of the genus *Canis,* while in *Vulpes* the dominant animal exhibits few changes in tail position or other posture.

Like other wild dogs the pampas fox has been hunted heavily by man, in part because of its reputation as a poultry-stealer, but mostly for its pelt. In Argentina the animal is known among furriers as the "zorro del país," or country fox. The decline in fox numbers has had an adverse affect on agriculture, as might be expected. Where foxes have decreased there has quickly been an increase in the numbers of field rats, viscachas, partridges and other enemies of crops, which cause much more damage than the occasional disappearance of a domestic fowl.

The Peruvian Fox *(Dusicyon inca)*

Angel Cabrera distinguishes the Peruvian fox, or "atok," as a separate species, though he admits that its exact relationship is in doubt, as the description is based on a single specimen. The animal's range is southern Peru, where it hunts at elevations as high as about 13,000 feet. Externally it resembles the pampas fox, except that the color is somewhat less vivid: the black back is grayer, the markings are tawny rather than reddish. The skull, however, approaches the dimensions of the culpeo, a race of which is found on the altiplano south of Peru.

THE CHILLA (*Dusicyon griseus*)

The chilla is the southern and western representative of the South American "gray foxes." In addition to its distinctive Chilean name of chilla it is also called the Argentine fox or simply the little gray fox. Its range is central Chile, the plains of western Argentina, and Patagonia all the way to the Straits of Magellan. The chilla presumably is more at home in forested terrain than the pampas fox, since in Chile it ranges as high as an elevation of 10,000 feet. Southern animals are larger and somewhat yellower than northern ones.

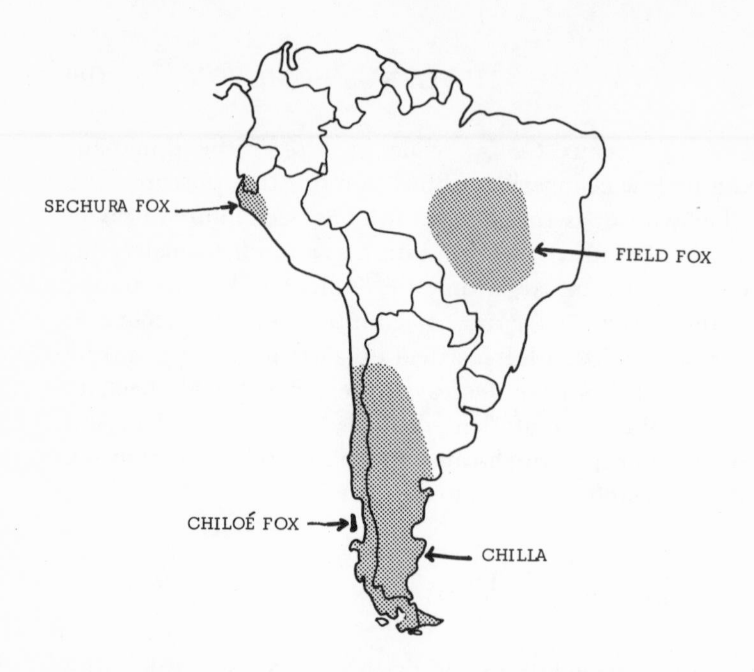

Chilla, the little gray fox of temperate South America. *Photo: Dr. M.W. Fox; copyright Jonathan Cape*

In length the chilla is only a little smaller than the average pampas fox, but it is a slender animal which gives the impression of being smaller than it is. The body averages 30 inches long, with the tail adding another 13 inches. The general coloring of the coat is similar to that of the pampas fox, but since the mixture of black and yellow on the back is more complete, the overall color is grayer in tone; likewise the black on the jaw and base of the muzzle is less intense, and the reddish-brown of the legs is much less vivid. The skull is somewhat different from that of the pampas fox; it looks lighter and more juvenile than that of the larger animal.

There is almost no information on the living habits of the chilla, other than that they are similar to those of the pampas fox—the chilla, for instance, also feeds on rodents found around cultivated areas.

The Chiloé Fox (*Dusicyon fulvipes*)

The small gray Chiloé fox, similar to the chilla in coloration, is found on Chiloe Island off the southcentral coast of Chile. Sometimes distinguished as a separate species, it really appears to be nothing more than an island form of the chilla of the mainland.

The Chiloé fox's principal claim to fauna fame rests on its encounter with Charles Darwin. During the round-the-world voyage of the *Beagle* the ship stopped off the southern end of Chiloé Island and Darwin went ashore with a crew of surveying officers. He spotted a fox sitting on some rocks, so absorbed in watching the surveying crew that Darwin was able to walk quietly up behind it and hit it on the head with his geological hammer. "This fox," he says, "more curious or more scientific, but less wise than the generality of his brethren, is now mounted in the museum of the Zoological Society." A similar lack of wariness in other members of the genus was noted by early settlers in South America. The trait must surely have helped account for the rapid extinction of the Falkland fox within a few years after the Islands were settled.

THE FIELD FOX (*Dusicyon vetulus*)

Considerably smaller than other members of the genus *Dusicyon,* the field fox is also called the small-toothed dog, the hoary fox, and in Brazil the "jaguapitanga." It is native to southcentral Brazil in the states of Goiaz, Minas Gerais, Mato Grosso, and the western part of Sao Paulo. The body measures 23 or 24 inches from nose tip to tail base; the tail itself is some 11 or 12 inches long. Adults weigh about eight or nine lbs. In addition to the smaller overall size, the proportions of the skull of the field fox are somewhat different from those of other members of the genus. The muzzle is relatively short, and the carnassial molars, both upper and lower, proportionately much smaller.

The coat is short because of the warm climate in which the field fox lives. Its color, like that of its close relatives, is a mixture of yellow and black, which results in a gray tone to the back and a lighter gray on the back of the head. The tail is the same color as the back, with a black tip and a black spot on the upper side near the base. The ears are reddish or tawny, becoming black near the tips; the point of the jaw is black but changes to whitish beyond the center of the jaw to blend with the white throat. The outsides of the legs share the red or tawny of the ears. Occasional animals range from darker-than-normal to almost black.

The field fox seems to occupy the same niche in its range as the pampas fox farther south. It is an animal of the open plains with a diet mainly of small rodents, birds, and insects, especially grasshoppers and locusts. It must also feed on vegetable food, since field foxes at the Rio de Janeiro Zoo eat bananas in addition to minced meat and raw eggs.

The natural history of the field fox is practically unknown. It normally has a small litter, not usually more than three, which is born in the spring. The Rio Zoo has had at least one litter of field foxes, four pups born in early September. Unfortunately they were abandoned by the female after the first day. The parents of this litter were kept together except during whelping and formed a harmonious pair, though once the male attacked the female and had to be separated from her for a short time.

Usually a timid animal, it stays away from man when possible, but is courageous in defending itself and its young. The Danish naturalist Lund recounts an incident in which he was forced to kill a female which in defense of her pups not only refused to retreat from him and his dogs but actually attacked them.

The Sechura Fox (*Dusicyon sechurae*)

The Sechura fox, which has the short muzzle and small carnassials remarkable in the field fox, is found more than two thousand miles from what seems to be its closest relative. It lives in the Sechura desert of extreme northwest Peru and in the arid part of southwestern Ecuador. This fox may represent a transitional form between the field fox and the other members of the genus, but only a great deal of further investigation will clarify its status.

Forest Fox
(*Cerdocyon thous*)

THE FOREST FOX is yet another South American wild dog with a baffling list of common names. Among others, it is called the savanna fox, the crab-eating fox, and Azara's fox. The name forest fox translates the Spanish *zorro de monte;* it is the equivalent of the Brazilian *cachorro do matto,* or that given the animal by the early naturalists Linnaeus and Buffon, *chien des bois,* both of which might be translated as "forest dog." Naturalists knew of the forest fox before they did any other South American canid, and many subspecies and local variants have been described.

The forest fox lives in the Guiano-Brazilian subregion, the part of South America covered by tropical or semitropical forests. This means that the range extends from Colombia and Venezuela in the north, south to Bolivia, northern Argentina, and Uruguay. Where this range is outside the areas of the great forests the forest fox is found in the clumps of woods dispersed on the savanna, or in the forests along the rivers.

A very unfoxy-looking animal, the forest fox is nevertheless quite elegant. The coat is shorter than that of animals farther south, and the ears are shorter too. The feet have non-stretchable membranes between the toes; the nails are relatively short and the pads under the toes and in the center of the soles are thick. The forehead bulges because of the large sinuses or frontal cells,

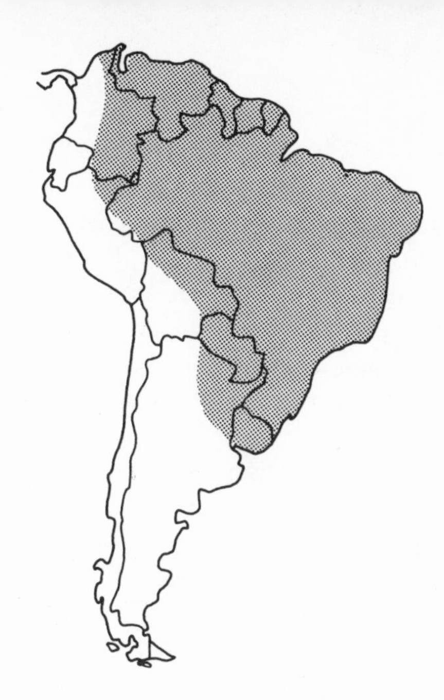

After Hildebrand (1954)

Forest foxes live in the forests and savannas of tropical South America.
Photo: San Diego Zoo

while the lower side of the jaw is markedly concave and broadly angled. The caecum or blind intestine, like that of the small-eared dog and the bush dog, is short and straight, rather than long and coiled as it is with most dogs. The forest fox is relatively small; the average length of head and body is 27 or 28 inches, while the tail adds another 11 or 12 inches. The average weight is about 15 lbs.

The brindled gray and pale bay color of the coat is caused by a mixture of yellowish-white and black hairs, with more black along the back and at the root of the tail. The face, the ears, and the fronts of the legs are reddish, while the throat and belly are pale or whitish. The end of the jaw, the tips of the ears, and the backs of the legs are black. There is a dark band across the chest, and the tip of the tail is black. Within this general pattern there is a great deal of variation among individuals; some are quite pale, others much darker, with nearly black legs and ears.

Despite the fact that the forest fox has long been known to science, little information is available on its habits. It seems to favor forested country and river basins but is also found in farmland and pastures and is seen on the outskirts of towns. It is normally nocturnal. Its diet consists of all sorts of small animals and birds; in Colombia, its principle prey is reported to be the cottontail rabbit. It also eats insects of all sorts, and, as one of its common names indicates, is reputed to be fond of crayfish and crabs. It is known to eat small turtles, and is also supposed to have a keen taste for poultry. In addition to animal food, however, the forest fox consumes a great deal of vegetable matter, such as fruits in season, seeds of various kinds, and cultivated crops—of which corn is a great favorite. Vegetable food may in fact make up the bulk of the fox's year-round diet.

The forest fox may be found singly, in pairs, or in family groups, but not in packs. Family groups of adult male and female and one or two half-grown young are frequently encountered. The Rio de Janeiro Zoo has kept some forest foxes in groups before dividing them into pairs and reports that a social hierarchy may develop, with weaker animals sometimes attacked by stronger ones, especially at feeding time. What little is known of their reproductive habits comes from zoos. London Zoo has two litters, numbering one and four pups, born in May and June. At the Rio Zoo a litter of three pups was born in late October but abandoned by the mother after she had nursed them for a couple of days. These pups were dark gray at birth.

A curious fact about the forest fox is its reputed tameability. In the Guianas, the Indians and Bush Negroes capture young foxes and raise them to hunt small prey. It has often been said that this fox can be crossed with the domestic dog to produce a hybrid especially useful in hunting. But, considering the presumed genetic distance between South American wild dogs and the domestic dog, such crosses are highly unlikely.

Small-eared Dog
(*Atelocynus microtis*)

ONE OF THE MOST DISTINCTIVE of all the South American wild dogs, and one of the rarest, is the small-eared dog or fox. This sleek, graceful animal ranges the rain forests of the Amazon basin in Brazil, Peru, Ecuador, and Colombia, and extends into the upper Rio Orinoco basin in Colombia and probably Venezuela. The animal's distinctive ears and unique coloring, its cranial peculiarities, and its unusual gait and posture have caused it to be placed in a genus by itself. We have no information at all on its life in the wild; what we know of its behavior is based only on descriptions of the animal in captivity. Although the small-eared dog is rare even in zoos, the Brookfield Zoo in Chicago has at one time or another kept several specimens.

To begin with, it is somewhat larger than most other South American wild dogs. It stands from 13 to 15 inches at the shoulder, with a head and body length of 28 to 40 inches. The tail is ten to 14 inches long—short in comparison to the body, but long enough to touch the ground when the animal is standing. The ears are only one and a half to two inches long: shorter in comparison to body length than those of any other member of the dog family. Two of the animals acquired by the Brookfield Zoo weighed about 20 lbs. each when they arrived at the zoo, but large specimens probably weigh considerably more. Like the bush dog and the forest fox, the small-eared dog has a short straight blind intestine rather than the usual long, coiled one.

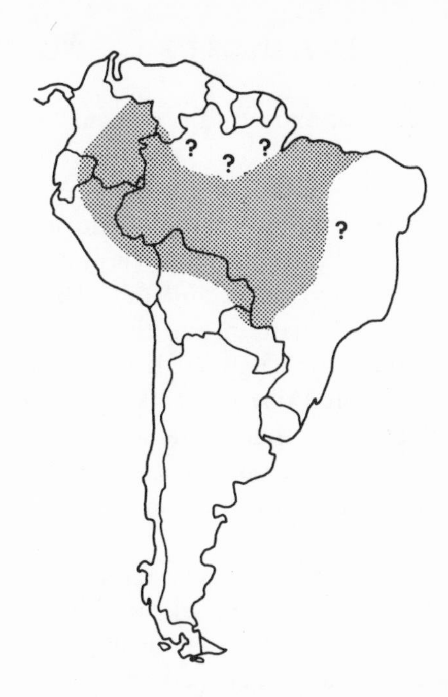

After Hershkovitz (1961)

Small-eared dog, well adapted to the tropical South American rain forest by its short sleek coat. ***Photo: Chicago Zoological Society***

The color pattern of the small-eared dog is another of its peculiarities, for it lacks the light underside common to most of the family. The coat is a grizzled brown or blackish, which shades inconspicuously into a dull reddish-brown on the underside. A narrow black "collar" is discernible, and a dark band runs along the top of the back and tail. The only light-colored area is a patch of buff hair around the anus and on the underside of the tail at the base. The coat is short, stiff and sleek, indicating that the small-eared dog is well adapted to a rainy climate or perhaps spends a lot of time in the water. The tail, however, is bushy and has a black tip, sometimes with a few white hairs interspersed. A band of hair along the top of the tail can be erected when the animal is excited.

The small-eared dog's posture is distinctive—when at ease, it stands with head lowered, forelegs slightly spread, hind feet with heels together and toes pointing out, and the tail held against the hind leg with the tip curved upward to keep it off the ground. Though its eyes are normally hazel-colored, in the light of a flashlight they shine with a bright, pale green intensity. The upper canine teeth are very long and protrude slightly beneath the lip when the animal's mouth is closed. Males exude a strong musky odor from the anal glands which gets more conspicuous when they are excited; observers at the Brookfield Zoo noted that the female did not have the odor. They also noted that the small-eared dog moves with a grace and lightness which seems almost cat-like.

The original pair of small-eared dogs kept at Brookfield lived together but unfortunately showed no attempts to mate. The male, though smaller than the female, was dominant in most activities. For instance, although the pair sometimes took turns eating from a common food dish, more often the male ate first, while the female snatched bits of food to eat at a distance or simply waited until the male was finished. Quarreling over food was accompanied by snapping, and by a snarling or snoring sound made with the mouth closed, but the animals never fought or bit each other. The pair used a common sleeping box, the male occupying the front while the female preferred the hidden portion in the back.

The two original Brookfield animals reacted quite differently to people. Never friendly, the female snarled in their presence. The male had been captured when very small and kept for some time in Bogotá, Colombia before being sent to Chicago. Although he had been hostile when captured, he gradually became tame and by the time he reached the Brookfield Zoo was noticeably friendly. He became eager for the attention of the keepers and regular visitors, and wanted to be petted and fed by hand. He responded by rolling over on his back and squealing with delight, and by a weak but unmistakeable wagging of the end of his tail.

After this engagingly friendly male died the Brookfield Zoo acquired another male to pair with the original female. Until someone undertakes the admittedly difficult task of studying the small-eared dog in the wild, anything else we may learn of the animal will have to come from such captive animals.

Maned Wolf

(Chrysocyon brachyurus)

THE BROKEN FOREST and savanna of central South America is the home of one of the most peculiar-looking of the wild dogs. Imagine a reddish animal with a sharply pointed muzzle, large pricked ears, longish hair raised in a mane on the back of a long neck, the habits of a fox, and the height of a Borzoi! This remarkable stilt-legged wild dog is the maned wolf or "lobo crinado," also called the red wolf, giant fox, and, by the Brazilians, "aguará guazú" or simply *"guará."*

The range of the maned wolf extends from the edge of the Amazon basin's rain forest in Brazil south through Paraguay and northern Argentina to the point where treed savanna is replaced by open plains. It may be found as far west as the eastern edge of Bolivia. It was once native to the interior of Uruguay but was eliminated there during the 19th century. It frequents the more remote lowlands and the densest brushland it can find—and avoids man to the best of its ability.

Part of its strange looks are due to the wolf's proportions. Its legs are so long that its height is greater than the length of its trunk. The total length of the head and body is 49 to 52 inches, and it stands about 29 inches at the shoulder. The tail, which falls only as far as the hocks at most, may be as short as 11 inches or as long as about 16.

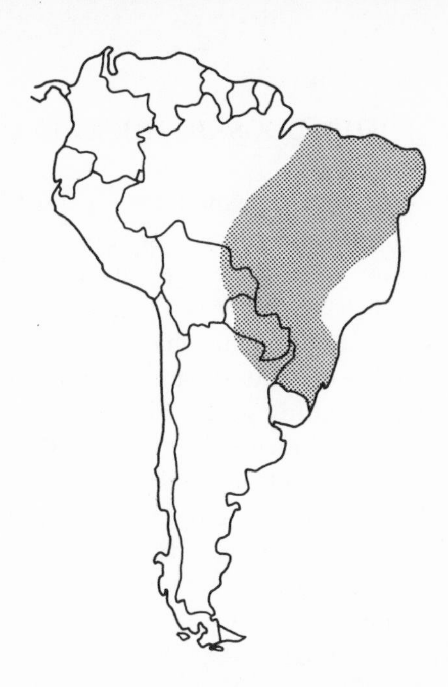

After Hildebrand (1954)

This maned wolf, photographed in Matto Grosso, Brazil, has just been run to exhaustion by horsemen. Its immensely long legs, short tail, and shaggy coat make it one of the most distinctive of wild dogs. *Photo: Sasha Siemel*

The coat of the maned wolf is fairly long and shaggy but it lacks a woolly undercoat. The longest hairs are on the back and nape of the neck, where they are raised to form a sort of mane. The animal's coloring is vivid and handsome, a chestnut red over most of the body, somewhat darker at the mane, and blending into black on the muzzle and on the legs and feet. The end half of the shaggy tail is white, as are the long hairs on the insides of the ears. It shares with the bush dog the peculiarity that the pads under the third and fourth toes are united to form one large pad.

Neither in habits nor in temperament is the maned wolf at all wolflike. It is a solitary animal, and one observer familiar with the animal in its native habitat reports that only once has he seen two together. In the Brasilia Zoo, keepers have found that two animals of the same sex may be kept together without undue quarreling, though individuals of opposite sex are put together only when the female is in heat. The maned wolf is gentle, even timid, and a newly caught animal may be touched and handled with impunity. The voice is a mournful "guaaa...!" which carries far at night, the time when it is most often heard. Captive maned wolves are very vocal, barking rapidly, howling with great intensity, and snarling at each other like cats when they quarrel.

Angel Cabrera vividly describes the field appearance of the maned wolf—"By moonlight especially he seems somewhat fantastic, trotting on his long legs, his head held low, the ears thrown back and the tail dangling, with rather disjointed movements, and appearing larger than he is by virtue of the raised hair on his back." Of no help to its ungainly appearance is the fact that the wolf's hind legs are slightly longer than the front. The animal seems to have a difficult time descending a steep slope and hunters take advantage of this by shooting it as it is going downhill. On the other hand, it can climb a slope with speed and agility.

The maned wolf seems to eat anything, animal or vegetable, which it can find and capture with its small teeth and weak jaws. It hunts rodents and other little animals, reptiles, ground birds and eggs, snails and slugs, frogs and cast-up fish. Its range of vegetable food is equally wide: bananas, guavas, sugar cane, and a great number of wild fruits. One of the staples of its diet is the fruit of *Solanum grandiflorum,* a relative of the potato, which in

Brazil is called wolf's fruit. An active animal, the maned wolf often travels long distances in search of food. Its diet at the Brasilia Zoo gives one an idea of the amount it requires; each animal is daily fed 400 grams of cooked rice, 150 of cooked oats, 100 of *Solanum grandiflorum,* 600 grams of bananas, 800 of cooked beef and, on alternate days, one chicken.

Like the foxes of the northern hemisphere, the maned wolf probably does its hunting and food-gathering by roaming back and forth over its territory, snapping up anything it comes across, and it would need to roam far and snap up a great deal to provide the equivalent of the Zoo's amount of food per day. One curious hunting technique reported of the maned wolf is its habit of using teeth rather than paws to dig out rodents. Probably its long, slender legs are not very good for digging, and anyone who has seen a terrier use its teeth to tear out roots and remove dirt from a rodent's hole will know how efficient a means of excavation this can be.

The maned wolf is popularly accused of supplementing its diet with domestic animals. Some people, however, deny that it kills even poultry, much less the sheep that it is reported to take. Though its overall size may be the chief basis for the accusation, the wolf's timid disposition, along with its weak jaws and teeth, would seem to absolve it from the accusations of sheepkilling.

Very little is known of the social and reproductive behavior of this animal in the wild, though specimens in captivity have provided us with some information. Females come into estrus between April and June and mating behavior resembles that of other members of the family. Both sexes are more active at this time, the female running about, lowering herself in front of the male, rubbing against him, and beating her fore feet rapidly against the ground; while the male follows her, smells and licks her vulva, and attempts to mount her. According to observations at the Brasilia and Frankfurt Zoos, estrus lasts for about five days and copulation includes the tie, which lasts for ten to 12 minutes but can quickly be discontinued if the animals are disturbed.

The gestation period appears to be about 65 days and litters are usually small. Though some sources state that there are only two pups per litter, Brasilia has recorded a litter of three, and Frankfurt Zoo a litter of four. In the wild the female bears and raises the pups in a nest made in the thickest and most secluded vegetation she can find. At birth the pups are a dark brownish gray and only develop their characteristic long legs when they are several months old.

Presumably the maned wolf is prey to the usual canine diseases and parasites. In captivity it has been noted that infestation with the kidney nematode *Dioctophyme renale* is frequent. This parasite is transmitted by the animal's eating infested fish. Infection results in itching and irritation of the skin with much shedding; tumors of the skin; regurgitation; lack of thirst; and urine retention. The Sao Paulo Zoo has lost at least five animals to this disease, which is fatal once infestation has occurred and can be controlled only by removal of the infected kidney.

The maned wolf probably has few natural enemies, but it falls an easy prey to man, and because it is neither cunning nor exceptionally fast, it can be lassoed from horseback. There is no reason why it should be hunted, for its coat, lacking an undercoat, is worthless to furriers, the animal is apparently harmless to livestock, and it is beneficial in its role of eater of insects and rodents. Yet skins of the maned wolf can be found for sale at pitifully low prices in the interior of South America. Based on the number of those captured in the last several years in Brazil, zoologists there estimate that there may be some 1500 to 2500 maned wolves living in that country. What populations may number elsewhere is not known.

Raccoon Dog

(Nyctereutes procyonoides)

ONE OF THE MOST curious and least familiar wild dogs is a small animal which is marked like the raccoon, hibernates during the winter, and though native to the Far East has been successfully introduced into eastern Europe. This is the raccoon dog. Its original range extended from eastern Siberia south through Manchuria and Korea, across mainland China as far west as Szechuan, and into northern Vietnam. It is native to Japan and Sakhalin Island, but not to Taiwan. In Japan, where it is called *tanuki,* the raccoon dog was once common to all the principal islands, and was hunted widely for its fur, but now—though it still figures prominently in Japanese folklore and superstitions—it is restricted to only a few places.

In 1928 the Russians, in an effort to capitalize on its valuable fur, began to introduce it into other parts of the U. S. S. R. Some 4000 animals were released in 40 different regions around the Soviet Union, and while in some places the raccoon dog did not thrive, in others, notably European Russia, it prospered and spread until now it has populated Finland, Poland, and parts of Sweden.

The size of a fox, the raccoon dog is far less svelte in its general appearance. Its head and body are from 20 to 32 inches long, but its tail, five to ten inches long, is unfoxlike in that it is usually well under one-third of the body length. The animal

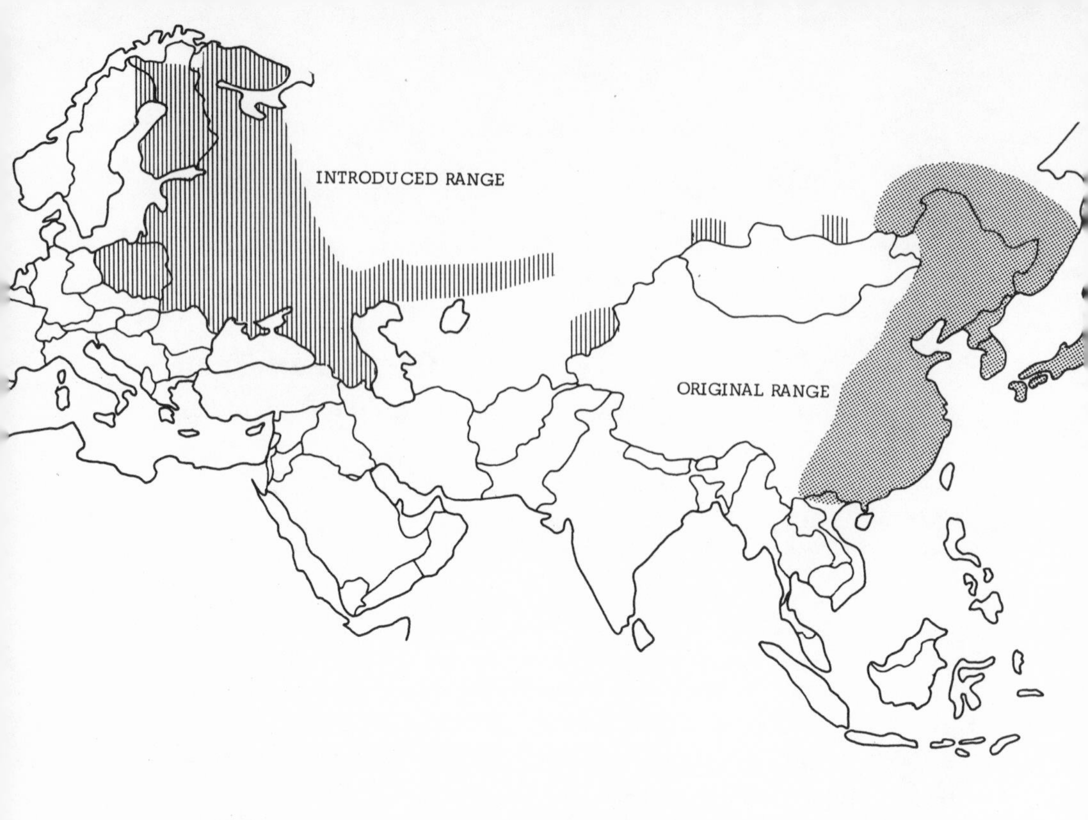

A raccoon dog in its native Japan. Because of its preference for thickly vegetated river banks and reed beds it is not often photographed in the wild.
Photo: Kojo Tanaka

stands about 15 inches at the shoulder and weighs from eight to 13 lbs. in the summer, though it gets as heavy as 20 lbs. in late fall and early winter as it prepares for winter sleep. The muzzle is fairly short but sharply pointed, the ears short and rounded at the tip. Its skull and dentition resemble those of most other dogs, except that it occasionally has an additional molar in the upper jaw.

The raccoon dog's general appearance is affected by the condition of its fur, which in cold weather is thick and soft underneath, with longer guard hairs. These outer hairs are particularly long on the sides of the head, which gives the impression of its being disproportionately broad, while the luxurious fur often hides all but the tips of the ears. Similarly, the long hair falls over the rump and covers the upper portion of the short tail so that the hindquarters look rounded. Heavy fur on the belly makes the legs seem shorter than they are. In its summer coat, the animal stands well up on its legs and exhibits alert, erect ears, but in winter, under its mass of fur, it seems to stand barely off the ground.

Brownish-gray over much of the body, it has a darker back—and a vague dark cross sometimes marks the shoulders. The belly and the insides of the legs are a yellowish-brown which sometimes extends up the throat, though in many animals the throat is a deep brownish-black, the same color as the outsides of the legs. The muzzle is dark, and the animal has dark rings around the eyes which form a raccoon-like mask. The tail is never ringed, though; it is yellowish-brown below and dark above.

The raccoon dog's preferred habitat provides dense protective cover that abounds with the small animals and vegetable matter the dog eats. It is thus found primarily in stream and river valleys where the climate is sufficiently damp to produce thick undergrowth. It rarely occurs in coniferous forests, presumably because such forests lack both the ground cover and the variety and amount of food the animal requires. It thrives in the marshes and reedbeds of river mouths, and reportedly is a fine swimmer.

Primarily nocturnal, the raccoon dog is about during the day when it is pressed by hunger. Out foraging, it walks at a characeristically quick, nervous pace, nose near the ground, in a constant search for anything edible. Its paw print is more rounded than that of a fox, and the trail forms a broken line, not a continuous one. Though it does not move very fast, the raccoon dog is highly mobile except during the breeding season; this habit of roaming restlessly in search of food is one of the traits which has allowed it to take over so rapidly the areas into which it has been introduced.

The animal retreats to a burrow during the day, but prefers to use one dug by a badger or a fox rather than to dig its own. If a burrow is not available the dog will build a nest or den in heaps of old straw or reeds found in marshy or swampy areas. It will also use crevices in rocks, rotten hollow logs, windfalls, rotted roots, even the old military excavations and trenches found all over eastern Europe and the western U. S. S. R. In Japan, it sometimes clambers into trees along sloping branches close to the ground, and may den in hollow trees or under the roofs or lofts of temples (it is quite common around inhabited areas).

Winter hibernation is the raccoon dog's most distinctive characteristic. It eats heavily in the fall, frequently increasing its normal weight by 50 percent. The winter sleep is triggered by cold, heavy freezing, or snow fall, and may begin as early as November and last into February. The raccoon dog retreats to its den or burrow, often lined with an effective insulation of dry grass or moss, and usually shared by pairs or family groups. Its sleep is not deep and it will awaken and forage for food if the weather turns warm. In the southern parts of its range it does not hibernate at all, and its winter sleep in the colder areas depends on its being able to store sufficient fat. Poorly nourished animals are forced to remain active all winter, even in the north.

The raccoon dog is truly omnivorous, though its small size limits its intake to small prey. The proportions of different kinds of food vary with area and season, but everywhere raccoon dogs depend on frogs, voles and mice, insects, reptiles, mollusks, fish, birds, fruit, berries, and grain crops.

In some places, under special conditions, they may concentrate on one type of food—fish during a spawning run, for instance. They are reported to be excellent fishermen, sitting beside streams to scoop out fish with a paw, raccoon fashion, or plunging into the water to swim and dive after their catch. Along the seashore they collect dead birds, fish, or any edible debris cast up by the waves; they also catch sea urchins and crabs. In some areas they prey heavily on game birds, and occasionally on domestic fowl.

Autumn with its abundance of grain, fruit, roots, and seeds, allows the raccoon dog to fatten up for the winter. During this season it still takes many mice, but eats fewer other animals such as frogs. Winter and early spring are difficult times, especially in the northern part of its range, or where poor conditions have kept it from fattening itself adequately during the fall. During winter, rodents still make up the bulk of the dog's diet but they are hidden under the snow and take time and precious energy to hunt. Under these conditions the non-hibernating raccoon dog is often reduced to eating horse manure and refuse around villages, and may even enter towns in search of food.

Although the raccoon dog is bred on fur farms in the Soviet Union, many details of its reproductive behavior in the wild are unknown. For instance, authorities are not quite certain whether the animal is polygamous or monogamous. The mating season is in the spring, heat beginning with the first warm days of February or March, or even at the end of January. Courting takes place sometimes between individual pairs and sometimes among one female and several suitors, which fight, not very savagely, among themselves. Pro-estrus is marked by increased socializing among adults; at times there is alternate scent-marking of objects, and both males and females, when they are separated, make a mewing or whining sound. The female holds her tail slightly raised to expose the vulva, which during the estrus period of about a week becomes swollen, as do the nipples.

Although sexual behavior in the raccoon dog is in most ways similar to that of other members of the family, it has a few peculiarities. When the male is sexually excited he holds his tail raised at the root and extending outward and downward in an inverted J-shape. This peculiar tail posture has been noted for only one other canid, the bat-eared fox. Although copulation includes a tie, in two observed instances at the London Zoo the male did not dismount from the female to assume the tail-to-tail position common to other dogs, but remained mounted while the female lay on her back, side, or belly throughout the ten minutes or so that copulation lasted. The average gestation period for the raccoon dog is 61 to 63 days. Litters are large, usually seven or eight pups but sometimes as many as 16 or even 19.

Young raccoon dogs are born blind, covered with soft blackish fur, and weighing two or three ounces. Their eyes begin to open on the ninth or tenth day and their teeth appear by about the 15th day. When the pups are still less than two weeks old, the guard hairs start to grow, first on hips and shoulders and then around the cheeks and ears. By the age of four to six weeks the fur is clearly divided into undercoat and guard hairs, though both layers are still short. The pups begin to emerge from the burrow when they are about 15 days old, and by the third or fourth week the parents are bringing them food, though they continue to nurse until they are six or eight weeks old.

When the pups are two and a half or three months old, at about mid-summer, they start going with their parents on hunting trips; before they are six months old they are self-supporting. By then they are almost the size of adults, though somewhat lighter in color. But pup independence does not necessarily signal the break-up of the family, for these dogs are social animals. Sexually mature between nine and 11 months, they mate at the end of their first year.

The raccoon dog needs its high fertility and adaptability, for it has a number of predators. It cannot run fast and therefore falls easy prey to wolves, its most dangerous natural enemy. It is prey too for wolverines, lynxes, yellow-throated martens, sometimes foxes, and occasionally such large birds as the golden eagle, sea eagle and eagle owl. (Its favorite habitat of densely vegetated river bottoms provides it not only with food but also with protective cover against its enemies.)

Man, however, is the chief source of danger to the raccoon dog. In the Soviet Union it is hunted for its valuable pelt, chiefly with gun and dogs, though sometimes with traps. Since its introduction into the northwestern part of the U.S.S.R. it has become one of the principal furbearers there. The fur is called "Ussurian raccoon" by the Russians and "Japanese fox" on the western market.

It was the commercial value of the animal's pelt that prompted the Russians to undertake their original acclimatization program, which has been notably successful in terms of the immigrants' viability. The raccoon dog did not thrive in the most northern and southern parts of European Russia, though it still exists in small numbers along the Arctic Ocean and in the Caucasus. Neither did it prosper in the Asian part of the country outside its original range. But in the central part of European Russia the raccoon dog took hold amazingly well, and started almost at once to expand its range. By 1955 it had reached Poland, had arrived in Finland even before that, and by now lives also in Sweden.

Since the raccoon dog likes the reed thickets of river bottoms in eastern Europe, we can predict that its future spread will be westward along the rivers into eastern Germany and eventually into western Europe. The spread of this small fur-bearer, like that of any alien animal, is a mixed blessing, of course. Its fur is valuable, but it is a disease carrier and has made a nuisance of itself by preying on eggs and fledglings in areas where many game birds breed. It might be particularly destructive in the great wildfowlbreeding marshes of eastern Europe. It will be interesting to follow the progress of this busy little dog as it moves across Europe, for inevitably it will attract more and more attention.

African Hunting Dog

(Lycaon pictus)

THE AFRICAN HUNTING DOG is an atypical member of the dog family. Thanks to its highly developed group organization, it is one of the world's most efficient predators. A diurnal hunter with little fear of man, its behavior can be observed with relative ease. Anatomically it has several distinctive features. Unlike other wild dogs, with their neutral tan, gray, or brindled coloring, the African hunting dog has a bright pied coat of black, white, and yellow. Fortunately, the African hunting dog's range includes the Serengeti Plain of Tanzania, where researchers in the past several years have added greatly to our knowledge of this formerly misunderstood and frequently despised carnivore.

Also known as the Cape hunting dog, the African wild dog, and the hyena dog, it is restricted to sub-Saharan Africa. Its preferred habitat is grassland, arid bush, and open wooded country. Within that habitat it ranges from South Africa north to Ethiopia and the Sudan, and west from the Ivory Coast and the eastern border of Guinea in the south through Mali and Niger to southern Algeria. Throughout its range the hunting dog is, along with the great cats, a major predator—the only one capable of pursuing plains game over extended distances.

Nowhere is it common; the extension of farming has restricted the areas where it is tolerated and has curtailed the amount of available game. In South Africa, for instance, it now occurs only in Kruger National Park and the parks of Zululand, and even in East Africa its numbers are dwindling. In West Africa the great decline in game has reduced its numbers severely.

The African hunting dog is a large animal, approximately the size of a German Shepherd. It stands from 24 to 30 inches at the shoulder and weighs from 60 to 80 lbs. Body length reaches 44 inches, with a tail of about 14 inches. Though its size is that of a large domestic dog, the proportions are different. The ears are erect, unusually large, and conspicuous for their oval or rounded tips. The animal's face and muzzle are short and broad and, together with the ears, give the head a proportionately heavy appearance. There are 42 teeth. The body is slender and wiry and the legs especially thin, though well muscled. It has a strong natural odor. The most noticeable anatomical peculiarity of the hunting dog is its lack of the fifth toe, the pollex or "dewclaw," on the front foot, which makes it unique among dogs. It has a strong natural odor.

Its pied coat is short and coarse, except for a mane of long hair under the throat, and a bushy tail. Young animals are almost completely black and white but as they mature, spots of yellowish-red appear, so that the adults are commonly three-toned. The end half of the tail is usually, though not always, white. There is much variation in the asymmetrical color patterns, with East African animals perhaps having less yellow than those farther south. Frequently the throat ruff, backs of ears, face and muzzle are dark, and there is often a central black line down the forehead between the ears.

The spotted coat is hard to explain. The dog's present mode of hunting, which involves an open approach to the prey followed by steady pursuit, does not require the mottled camouflage pattern found in the stalking cats. However, the dog's sophisticated method of group hunting is certainly the end of a long evolutionary process. Originally, it may have lain in wait for its prey, well disguised by its shape-concealing blotches.

The African hunting dog seems to have at least three distinct calls. There is an alarm bark—short, deep, hoarse, and sometimes accompanied by growls. Then there is the soft communal call or howl—a sort of *hoo-hoo-hoo*—which pack members use to locate each other. The third sound is a bird-like whimpering or chatter-

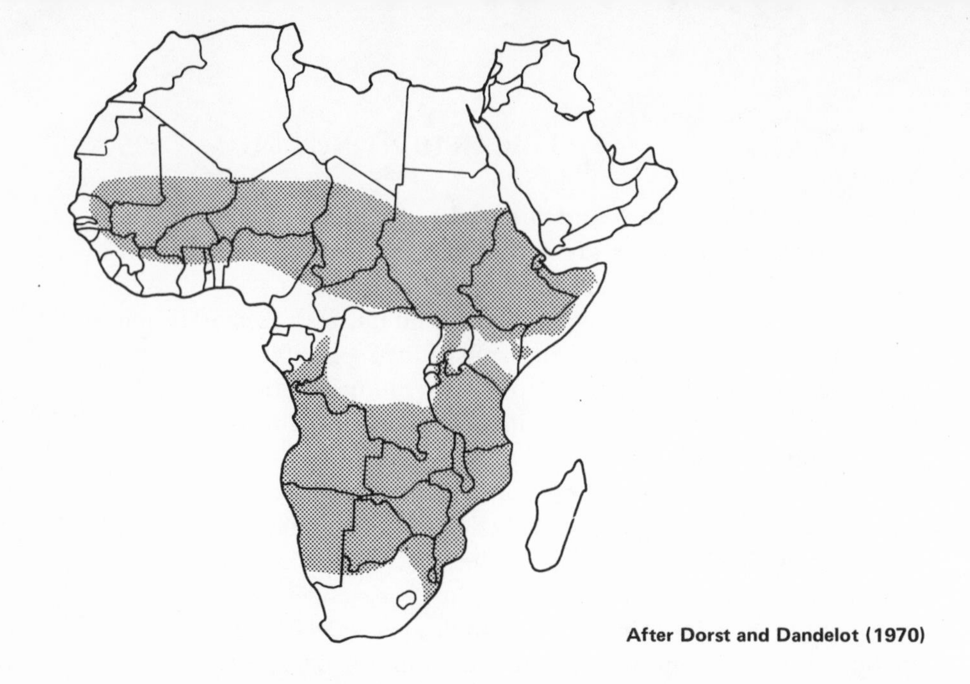

After Dorst and Dandelot (1970)

African hunting dogs live on open plains and treed savannas and are usually seen in groups. *Photo: Robert R. Wright*

ing often heard when the dogs are playing. It is hard to imagine anything less dog-like than this sound, which resembles the noise of monkeys.

African hunting dogs are sometimes seen singly or in pairs, but the common unit of organization is the pack, which usually consists of several adult animals and accompanying pups. Groups of from ten to as many as 40 animals are common. Reports of packs as large as 100, especially in southern Africa, may reflect both the exaggeration of the observer and the fact that strange packs of hunting dogs mingle amicably and may associate for a short while. Although some packs are polygamous groupings of one male and several females, others have many more adults males than female. Exactly how a pack develops is not yet known, but since social relations do not depend on a system of dominance, group organizations may be quite random. To be successful, a pack has to contain at least some good hunters, however, and one or more females capable of bearing and raising pups.

This association in large packs, as well as the open terrain over which they hunt, has helped make the hunting dog an unusually efficient predator. Early European observers in Africa were impressed above all by the seeming inevitability with which the hunting dog made its kill. Remarks are on record that no animal could hope to escape once these predators had set out after it. This is not completely true: hunting dogs, like other predators, watch and test their prey and sometimes give up a hopeless chase. But hunting efficiency in a given pack may sometimes be as high as 85 percent, and the average time to make a kill can be as little as half an hour.

Another fact about hunting dogs which has impressed observers is the respect in which they are held by plains game. When reading about hunting dogs one sees repeatedly the assertion that all plains game flees when these animals are in the vicinity, and that no predator is more dreaded by the herbivores of Africa. The respected ethologist Niko Tinbergen says, for instance, "No predator causes such a widespread panic among the plains game as these dogs; all round them the herds of Tommies run away with curious stiff jumps. . . . Waves of such panics move over the plains, and one can spot hunting dogs from several miles away by watching these 'dreads.' "

So many people have noticed this behavior on the part of prospective prey that it certainly must occur. But it is far from inevitable. Game animals often know whether or not the dogs are hunting. When they are approaching prey in earnest, they assume a characteristic head-low position and this changed silhouette may have something to do with the reactions of their prey. Many recent observers tell of seeing prey animals stand calmly or even approach hunting dogs, and Wolfdietrich Kühme, whose extended observations of hunting dogs in the Serengeti have contributed greatly to knowledge of the animal, notes that in the four months during which he watched hunting dogs there was no sign of alarm on the part of prey animals.

African hunting dogs certainly have highly stereotyped hunting patterns. The daily routine, involving a hunt in the morning and another in late afternoon, appears to be unvarying. One pack which Kühme studied consisted of eight adults and 15 pups. An antelope weighing some 65 lbs. was killed each morning and evening while the pups were small, while the kill increased to two animals each morning and evening as the pups grew larger.

A typical hunting incident begins when a member of a resting or sleeping pack arises, stretches, yawns, then walks over to another dog and greets it—a greeting which is passed from animal to animal until soon all are up and playing together. Eventually the pack, with the exception of young pups and adult babysitters, moves out to the hunt. Sometimes there is a definite order, a certain animal leading, with other specific ones in second and third place, though occasionally no such order can be discerned. But in any pack experienced adults are in the vanguard, with younger animals, often more playful than their elders, bringing up the rear. Sometimes the pack moves in single file; at other times it spreads out. Movement is at a slow trot, the animals searching for prey more with eyes than ears or nose—occasionally jumping high to peer over the tall savanna grass.

When potential prey is spotted, perhaps more than a mile away, the dogs move toward it, still at a slow trot, head and ears raised. Once the distance has been reduced to 500 yards or less, the pack slows down and falls into single file. As they move at a slow walk straight toward the target, their heads are held low, in line with the body, and ears are bent back. By this time the prospective prey has usually seen the dogs, watched them approach, and turns to flee—sometimes when they are as much as half a mile away but often only after they have approached to within 200 yards or less.

When the prey wheels and runs, the dogs break into a run in pursuit. This appears to be the usual sequence—that is, it is the flight of the prey which usually triggers the rush of the predators. Ordinarily the prey consists of a herd of small or medium-sized antelope; one particularly nervous antelope first breaks into a run and the others promptly follow.

The hunting dogs dash in pursuit, initially chasing different animals until one dog, usually the leader, gets closest to his target. Then all the dogs turn to concentrate on chasing this one slow antelope. Unless confused by the escaping animal's sudden mingling with another herd, the dogs stick to just this target, no matter how close they pass to other game. Such a chase rarely lasts more than four or five miles, often quite a bit less. Hunting dogs can run between 30 and 40 miles an hour over long distances. Antelope can sprint faster than this over the short haul but lack the stamina to keep up such speeds.

Hunting dogs in pursuit of game run in more or less single file, the lead dog as close to the prey as possible, with the second and perhaps third dogs at intervals behind and the rest of the pack loping along in the rear. If the hunt is short this order does not usually change, and the lead dog is the one that catches up with the antelope, bites it, and pulls it to the ground. If the chase is long, however, the lead dog may tire and drop back, so that the second dog moves up to take its place. As the antelope tires it starts to zigzag, and the laggard members of the dog pack simply cut off corners and prevent this final desperate tactic from succeeding.

The method of killing, like that of the hunt, is stereotyped. If the prey is small, the closing dog leaps to clamp its jaws into the soft part of the antelope's flank, causing it to turn and fall. The dog then bites into the belly and pulls out the intestines. A larger antelope may not be brought down so quickly, so the leading dog or dogs bite and slash at hindquarters and flanks until the prey is weak or stiff enough to be pulled down. No matter how large the pack, only a few dogs—those leading the group—do the actual killing. There seems to be no correlation between the size of the prey and the size of the pack. What function the trailing members of the pack have in the hunt is not clear, unless it is to cut off any final zigzagging of the prey, assuring that the lead dogs' energy is not wasted and that almost always, when there is a chase, there will be a kill.

Hunting dogs are said to kill nearly every sort of game animal except elephants and hippos. They have been seen harrying a solitary lion, and indeed, an old, weak one might fall to them. They can bring down such large animals as greater kudu and waterbuck, and hartebeest and wildebeest are not infrequent prey. When small antelope are available, however, as in the Serengeti with its untold numbers of Thomson's gazelles, and in other regions which abound in impala and Grant's gazelles these small antelope make up the bulk of their diet.

Although the success rate of the hunting dog is very high *once the pursuit has begun,* it does not mean that African plains game lacks defense against a dog pack. A victim can run for a river or waterhole and plunge in; hunting dogs reportedly will not follow, perhaps because they are afraid of crocodiles. Or an antelope can run through neighboring herds of grazing animals, shaking off its pursuers in the confusion. But the best defense is to insure that no chase takes place. Since it is flight on the part of the prey which usually triggers pursuit on the part of the hunting dogs, a herd which refuses to run is usually safe, and wildebeest and zebras especially put this fact to good advantage.

Hunting dogs approaching a wildebeest herd usually have their eyes on the calves. When the dogs are spotted, cows with calves retreat to the rear of the herd, while the remaining adults form a line and advance slowly on the approaching predators. As the dogs attempt to outflank the skirmish line, the wildebeest wheel and maneuver to keep themselves between dogs and prospective victims. Sometimes the dogs and the adult wildebeest get within a few feet of each other, the dogs making no attempt to attack. When cows and calves have retreated a safe distance the remaining wildebeest move off after them—and the dogs appear to lose interest. If a nervous calf suddenly breaks through the protective line during maneuvers, however, it is pulled down in an instant.

Once a kill has been made, dogs already on the scene start to feed immediately, tearing large chunks from the carcass with their powerful jaws. As other dogs arrive they too begin to eat at once, yet there are no conflicts over sharing the kill. Sometimes the trailing dogs are so far behind the leaders that when they reach a kill, particularly if it is a small antelope, there is nothing left. These new arrivals do not go hungry, however; pups and even other adults beg food from animals that have already fed, and food is regurgitated for them. The same meat may enter several stomachs before it is actually digested. This system ensures that all members of the pack, young, old, and crippled, share the kill.

The hunting dog has sometimes been accused of killing more than it can eat, but this does not often happen. The packs are large enough, and individual eating capacity great enough, so that, unless they are disturbed, hunting dogs almost invariably strip a carcass clean.

The unique social organization and communal feeding of African hunting dogs seem to have evolved from patterns developed for the care of the young. Breeding takes place throughout the year, with a peak from March to July. Estrus appears to last from three days to a week or more. At the Nairobi Zoo, when one female of the pack of three females and four males at the zoo came in heat, one of the males followed her about, resting his head on her flank when she stood still or lay down. The other animals paid no attention to the pair and there was no fighting among the males.

Copulation, which took place several times a day over a two-day period, was brief, involved relatively little thrusting compared with domestic dogs, and did not include a tie. If the lack of a tie is typical, it points to a considerable difference in behavior between the hunting dog and other members of the dog family.

Nairobi, Amsterdam, and Pretoria zoos have all recorded gestation periods of from 69 to 73 days or more. Before a litter is born the adults select a den site, usually the abandoned burrow of a warthog or aardvark. Typically, the dogs' denning area will have several such burrows, in which all the pregnant females in the pack bear their young. This is the only time when the pack remains in one place. During the rest of the year hunting dogs range widely (one pack has been seen in locations 70 miles apart), and never stay put for long. But while the pups are young the life of the entire pack centers around the den.

Since several females may litter close together or in the same den, burrow counts of pups do not necessarily correspond to litter sizes; however, Kühme observed one female with 11 young, while the Pretoria Zoo reports litters of from two to 12 with an average of about seven. Pups are raised communally, any lactating female nursing any of the young.

The young are dependent on adults for a long time. Although they begin receiving regurgitated meat at two weeks, they stay in the den and continue to nurse until they are three months old. A litter fostered by a domestic dog at the Amsterdam Zoo provides some information on their early development. The pups were

African hunting dogs, the only spotted wild dogs, raise their large litters communally. *Photo: George B. Schaller*

black and white when born, with no hint of yellowish or reddish fur, and weighed 12 or 13 ounces. At 14 days the eyes opened, spots of pale brown were appearing in the black parts of the fur, and weights ranged from one-and-a-half to two-and-a-quarter lbs. At three weeks, the pups started clambering around and began to cut their teeth. As they grew older the pale brown spots on the fur gradually turned to yellow. The pups, at five weeks, weighed four-and-a-half to five-and-a-half lbs.; there was no way to record their weights after that age because they wiggled too much on the scales.

At three months, pups are able to leave the burrow and can be weaned, but they are not strong enough to travel with the pack at its normal speed and for the usual long distances. So they must continue to be fed regurgitated food and, with their guardians, stay near the burrows until they are at least seven months old. Pups as old as five months still take precedence when it comes to food, and they beg quite aggressively if hungry. Food-begging also involves a stereotyped, instinctive action: the pup pushes its muzzle into the corner of the adult's mouth to trigger regurgitation. An extraordinarily hungry pup may forget itself and go so far as to bite at the lips or legs of the adult, but it is the nose-thrusting against its elder's mouth which is the key gesture.

Other young animals among members of the dog family have similar routines for soliciting regurgitation. What makes the feature unique among hunting dogs is that the gesture has come to be used among adults as well—not solely to distribute food but to suppress aggression with appeasement. Hungry adults—those left with the pups as nurses or guardians, or those late in arriving at the kill—beg from successful hunters. Even over an antelope carcass, a hungry dog, instead of disputing another's place, begs from the feeding animal, and thus the entire pack may be satisfied without showing a sign of hostility.

The begging gesture is completely ritualized, and empty of feeding significance, in the adult greeting ceremony of the hunting dog. Especially during the socializing which occurs before the hunt, the adults mingle happily, jumping about, calling in their peculiar chirping voices, and assuming begging postures similar to those used during feeding. This greeting ceremony is the center of the animals' social system; it is a method of suppressing aggression and of achieving what seems to be total parity among adults.

Unlike other social animals, hunting dogs give no indication of a dominance order, nor are there any facial expressions for communicating threats or superiority. Although there are leaders among them when the pack is hunting, and although this order is frequently constant, hunting leadership carries no special rights and seems to have nothing to do with dominance.

Adult social life is egalitarian but such equality does not extend to the young. Pups have priority when it comes to begging for food, and at the kill itself adults will immediately retreat from the carcass when hungry youngsters approach or bite them away. But pups that approach adults outside of normal begging times, or without intending to beg, are frequently bitten away by the adults. Gradually, of course, older pups come to be accepted into the company of adults and the mutual adult greeting ceremonies can take place without friction.

The pack organization of hunting dogs makes them safe from the large cats, and though dogs and a solitary leopard or lion may contend for a carcass, their relationship seems generally to be one of mutual respect. Hunting dogs and hyenas have frequent contact with each other, and there are records of each driving the other off its kill. Although dogs have been seen to chase and torment a lone hyena, the two species seem frequently to ignore each other or merely to react with mutual curiosity. One observer actually reports having seen a hyena running with a dog pack, acting as though it were one of them.

With no natural enemies to keep them in check, what holds down the numbers of this prolific, unusually well adapted, and highly efficient predator? For one thing, of course, the hunting dog's predation efficiency can only be high where conditions are right—this means open plains country well-stocked with small and medium-sized antelope. Such favorable conditions never did occur throughout Africa and are harder and harder to come by on the modern continent. But one important population control may be disease. Distemper inflicts a high mortality on young hunting dogs, and several litters living close together or mingling in the same burrow are particularly susceptible to the rapid spread of disease. Accidents must also be frequent. Lame hunting dogs are often seen, and although they are fed by the group, they inevitably reduce pack efficiency.

At present the chief danger to hunting dogs is the violent hatred of man. Both European and African stockmen and farmers have always loathed it as a sometime killer of sheep and cattle. Cattle herders can drive dogs off if they can intervene quickly enough, for the dogs are not aggressive towards man, but if the cattle herd is large, spread out, or poorly guarded, the dog pack is in, has made its kill, and is feeding before the stockman can protect his animals.

Unfortunately the hatred of the African hunting dog is too often shared by those interested in preserving Africa's fauna. Hunters have traditionally shot the hunting dog on sight as vermin. And park rangers, who should know better, have in places embarked on extermination campaigns against it. The grounds for the uncontrolled killing of this predator are the usual ones given by those who insist that herbivores are best protected and their numbers fostered by wiping out their natural predators. But the antagonism against the hunting dog is compounded by what many people see as their ferocious, "inhumane" killing methods.

It is true that the hunting dog runs its prey to exhaustion, and that, unable to deal a single killing blow to a large antelope, it bites and tears repeatedly until the prey succumbs to shock or to a lethal wound. This is the same hunting method used by wolves, dholes and other pack-hunting dogs. Fortunately, recent work on the hunting dog has shown how its predation, far from decimating an area of game, fits the overall ecological balance of the African plains. It suggests that African herbivores may have evolved as they have in part to meet the threat of hunting dog predation. Since such predation is directed against the slower members of a game herd, it causes some movement among those animals—and may help to lower the rate of parasite reinfestation which rises as herbivores graze long in a single area.

New information also gives the lie to the hunting dog's reputation as a ferocious animal. It has never been a threat to man himself. There seem to be no confirmable reports of any attacks on people, though the curiosity of the animals does cause them to approach and circle travelers. Within its own species the hunting dog is socially one of the least aggressive, most highly sophisticated carnivores known. There is much yet to be learned about the African hunting dog—about its patterns of reproductive behavior, for instance, and more about its population dynamics. As study continues and these aspects are investigated, we hope that people will learn to admire this unusual African predator.

Dhole

(Cuon alpinus)

THE DHOLE IS ONE of those rather mysterious animals with an exotic name one sees often referred to in books on Asian hunting or natural history. Also called the Indian or Asian wild dog or hunting dog, the red wolf, and the red dog, it is as little studied as it is frequently mentioned.

The northern limit of the dhole's range is the U.S.S.R., where it occurs widely but is everywhere rare. It is found from the eastern Pamirs east through the Tien Shan mountains, Dzungaria, the southern Altais and the southern slopes of the Sayan and Stanovoi ranges of Siberia. It is found in the Amur basin and in Manchuria. From these northern limits the range extends south throughout India, but not to Ceylon. Though the dhole occurs throughout southeast Asia, including Burma, Thailand, Laos, Cambodia, and Vietnam, down the Malay peninsula and on the islands of Sumatra and Java, it is not found in Borneo. Now extinct in northeast China, it may still occur from Fukien Province south.

Except for its rounded ear tips, the dhole looks very dog-like. It is about the size of a small wolf, though both size and color vary considerably in different parts of its range. The head and body may be as short as 30 inches or as long as 44, and the length of the tail varies from a foot to as much as 20 inches. For its size, it has rather short legs, and stands only from 17 to 20

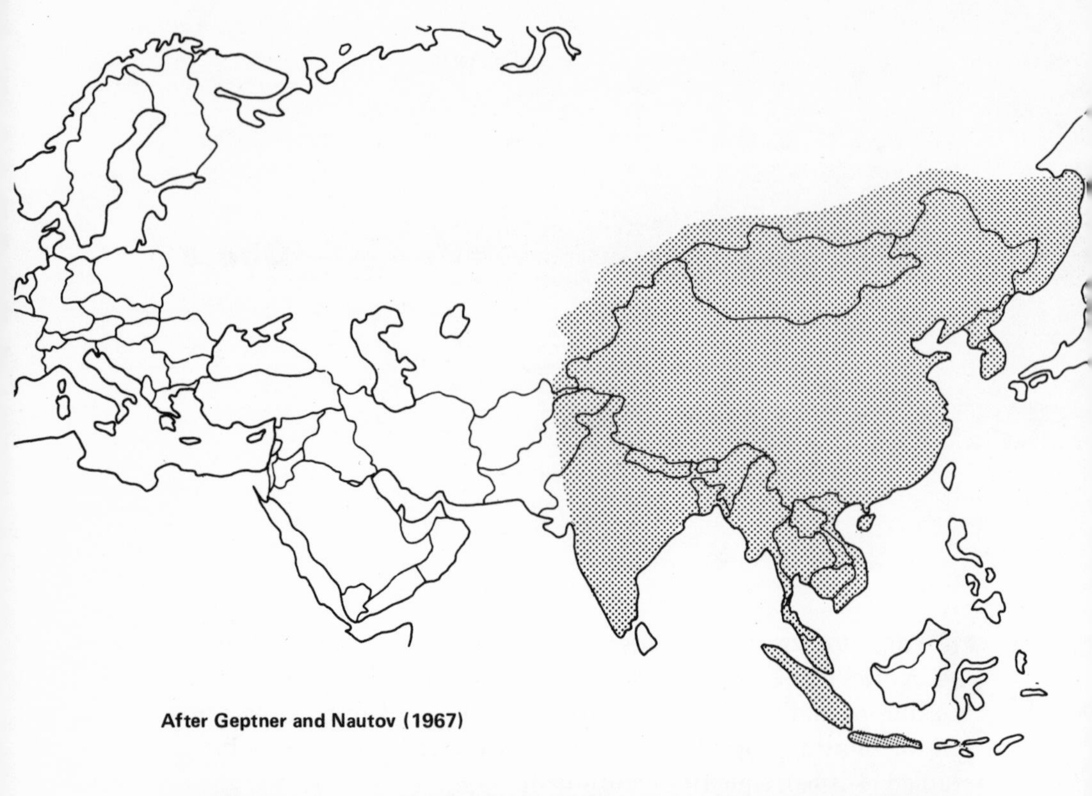

A dhole, one of a pack in Tamil Nadu State, India, pauses curiously to look at the photographer. *Photo: George B. Schaller*

inches high at the shoulder. The weight of a small or thin animal may be as little as 22 lbs. but larger ones average 35 or 40 lbs.

The dhole is easy to distinguish from wolves and jackals by its bright color and by its relatively full tail, which in winter fills out like a fox's brush and nearly touches the ground. The hair on the upper part of the body is a bright red, becoming duller—with a brownish or yellowish tint—on the top of the head, neck, and shoulders. The underparts, including throat, chin, belly, and insides of the legs, are sometimes lighter than the back, and may even be a creamy white; in other cases they are as reddish as the back. The tail is usually darker than the body and often ends in brown or black, though sometimes it has a white tip. Paws and lower legs also may be white. The face below the eyes and on the top of the muzzle may have more or less pronounced dark markings.

The brightness of the coat, as well as the vividness of the contrast between the back and the underparts, depends on the length of the coat; the more southern races, with shorter coats, tend to be more vividly colored. Coat length also varies considerably. Northern races have a very thick, woolly undercoat and long guard hairs in the winter, with thick wool on the insides of the ears that protrudes in tufts; long, thick side whiskers on the edges of the lower jaw, and hair covering the claws. Southern animals are shorthaired and sleek the year around.

Though the dhole looks much like a wolf or jackal, anatomically it has several peculiarites. Externally the greatest oddity is the six or seven pairs of mammae, even eight, instead of the four or five pairs found in wolves or domestic dogs. Features of the skull and teeth, when compared with the genus *Canis,* are distinctive. The dhole has only 40 teeth, the bones of the muzzle are relatively broad, and the skull has a convex profile.

The dhole prefers to live in heavily forested mountain country, up to elevations of 13,000 or more feet. In parts of its range, such as Tibet, where there are no forests, it inhabits the high steppes. In populated areas it sometimes hunts at night but by preference moves about in the morning and early evening. Dholes at the Moscow Zoo are very active at these times, and do a great deal of running, playing, jumping about, and standing on their hind legs—fore paws on the bars—to watch the world go by.

The dhole is a strong, agile animal for its size. A male in an enclosure at the Moscow Zoo managed to escape its pen by making a single running jump over a series of ditches, walls, and fences that totaled about 18 feet wide and at least four and a half feet high.

The dhole makes a variety of sounds which have not been studied in the wild, but animals at the Moscow Zoo are reported to whine, yelp, or chatter like excited magpies when fighting. Sometimes they make strange howling, whistling, hissing sounds. Although the dholes at this zoo live next to wolves and golden jackals they make no attempt to join these two species in their common howling sessions. The whistling sound may be an alarm signal. It can be imitated by blowing across the empty end of a cartridge, a means used by hunters in India to call the animal within shooting range. The dhole appears to run its prey silently, at least until the game is sighted, at which point the dogs may break into yaps and howls of excitement. When a howl or yapping is heard at night it may be a way of the pack's maintaining contact.

The pack is basically a family unit of the parents and young, but dholes are sufficiently social so that several families may be found together, and packs of as many as twenty animals are not uncommon. They prey on large hoofed animals; in India this means deer, wild pig, buffalo, and wild goats; in southeast Asia, it is deer, gaur (a large kind of wild cattle), and banteng; in Siberia, deer, wild sheep, and caribou. They also catch smaller mammals such as badgers and take birds.

In order to kill such formidable animals as the Indian buffalo and gaur, a high degree of pack organization is necessary, though single dholes tackle animals as large as the sambhar, a medium-sized deer. Like other wild dogs, dholes rely on their endurance rather than on great speed to tire their prey and bring it to bay. They can keep up a steady trot for many miles, and then close in on the exhausted quarry, some distracting it by biting at the nose and head while others cling to the flanks or rip at the belly. In parts of its range where the prey animals are migratory the dhole moves with the game; in the Sayan mountains of Siberia it is found up high during the summer and moves down to the lowlands in winter to escape the snow and to follow the caribou.

Much has been written of the dhole's ferocity, and it is often said that a pack of them will attack bears, leopards, and even tigers. They are declared capable of driving a leopard off its kill and have been seen harrying a sloth bear. But dholes probably do not systematically come in violent conflict with other large carnivores. The dhole is not accused of attacking man, despite its presence in a heavily populated part of the world. R. I. Pocock mentions the account of a hunter who shot two members of a pack and found himself surrounded by the remaining animals, which could have attacked him but showed absolutely no inclination to do so.

In the southern part of its range it may breed at any time of the year, though most of the young are born in January or February. Farther north, spring is the normal whelping time, with litters born from February through April. During estrus the male makes squealing sounds while pursuing the female, and copulation includes a tie which last from 15 to 20 minutes. The gestation period is probably about that of domestic dogs; at Moscow Zoo several litters have been born after an estimated 60 to 62 days.

The number of pups is usually from two to six, the Moscow Zoo having averaged four to five, but at least one litter of seven has been observed and an animal with up to 16 mammae would be capable of suckling even more. The dhole dens during the breeding season in a cave, a crack in the rocks, or a burrow, and several pregnant females from one pack sometimes den close together. The male aids in feeding the female and pups during the first weeks, and both parents regurgitate food for the weanling and growing young.

Born blind, dhole pups are covered with a thick, fluffy coat of brownish-yellow and gray fur, and have brownish-gray tails with white tips. They weigh from seven to 12 ounces. They grow very rapidly, nearly tripling their weight in the first ten days, and by two weeks the eyes are beginning to open and the teeth to appear. At a month, they begin to move about away from the mother, but they are still rather weak, and their actions are timid and cautious. They begin to nibble small amounts of meat at this age. Six weeks later they are much stronger, and run and play confidently. At two months they can readily gnaw meat off bones.

Although little is known about their social organization, it may be based on a system of dominance something like that of wolves. Animals at the Moscow Zoo often fight with others of the same sex in disputes over social position, but other members of the pack seem to pay little attention to the fighting animals. Perhaps they lack the intense interest shown by each member of a wolf pack in the social interactions of every other member. But too much cannot be inferred from such slim evidence. Perhaps one day enough will be known about the animal to compare it with the two very different systems among gray wolves and African hunting dogs, the other great pack hunters among wild dogs.

Bush Dog

(Speothos venaticus)

THE BUSH DOG of South America is a peculiar little creature, built rather like a heavy dachshund, which does not resemble at all the svelte foxes and wolves of the western hemisphere. The general range of this genuinely rare animal, called in Spanish the *zorro vinagre, perro de monte* or *perro de agua,* is northern and central South America. It is found from the eastern edge of Panama to the southern part of Brazil, and from the Guianas to the Andean Cordillera. It probably occurs west of the Cordillera Oriental in Colombia and Ecuador.

Built long and low, the bush dog looks rather like an otter or a badger. The body averages 25 to 27 inches in length, but the animal stands only ten to 12 inches at the shoulder. The well-furred tail is proportionally short—some five or six inches long. The hair is coarse and sparse; body color is dark grayish-brown, and (like the small-eared dog) it is peculiar in being black rather than light on the belly and lower parts. The relatively large, snub-nosed head, the small ears—about an inch and a half long—and the shoulders are a lighter reddish or tawny.

Anatomically, the bush dog is as peculiar inside as out. The total number of teeth is only 38, and the caecum, or blind intestine, is short and straight (a characteristic the bush dog shares with the forest fox and the small-eared dog) rather than being long and coiled as it is in most dogs.

The bush dog is usually found in relatively open country where savanna is interspersed with forest. It dens, evidently the

The bush dog, native of tropical South America, hunts in packs and swims and dives with ease. *Photo: San Diego Zoo*

year round, in burrows. It has some peculiar habits—for instance, it lives near the water by preference and likes to swim; not only that, it can even dive and swim under water very well. Its voice too is remarkable; it makes a range of clicking, whistling, chirping noises as it moves about while hunting and can produce piercing squeals and cries when excited.

What is known about the bush dog's hunting behavior comes from local informants; there seem to be no firsthand reports by naturalists of its behavior in the wild. It is said to hunt collectively in packs of up to a dozen, bringing down animals as big as deer, and preys especially on the paca and capybara, large South American rodents which take to the water when pursued. When hunting them, a part of the bush dog pack reportedly pursues the rodent on land, and when the prey takes to the water, the rest of the pack is swimming there waiting for it. Like most other wild dogs, the bush dog picks up small animals and vegetable matter found along the waterways where it lives.

The very little we know about social and reproductive behavior is based on a few zoo observations. Bush dogs can be kept successfully in small groups, and within the sexes show definite dominance patterns. Pre-copulation behavior is similar to that of other dogs. During pro-estrus the female's genital area becomes swollen and the male follows her, licking and smelling. She in turn increases her scent-marking activities and presents her hindquarters to the male. Actual copulation differs from that of most dogs, however, since it is questionable whether a tie occurs, and the back-to-back position which most dogs assume after the tie has never been observed.

The length of gestation is also in question. A pair of bush dogs at the Lincoln Park Zoo in Chicago were observed copulating repeatedly over a period of four days. A litter of three pups was born by Caesarian section 76 days from the last observed mating and 83 days from the first observed mating. This indicates a long gestation period for a dog, but even so the pups seem to have been somewhat premature.

There are reports that bush dogs are sometimes captured by Indians and raised as pets. Marston Bates was sold a female by a local woman in Colombia who had raised it from a pup. She had found it very dog-like in intelligence and behavior. Icty, as Bates named her, greatly enjoyed attention from people, always, he says, "greeting those of us who played with her with piercing

Bush dog pups are easy to tame and grow up to act very much like domestic dogs. *Photo: San Diego Zoo*

squeals and excited tail-wagging. She loved to have her ears 'scruffled' and would roll over on her back for a bellyrub, nipping our shoes or pulling our trouser legs if our attention lagged." This squeal, which she uttered with ears plastered tight against her head, was short, high-pitched and positively deafening.

Her recognition or attention call, given when people left her, when they approached, or when she wanted her food, was different: a series of short, metallic barks, "ungreased" as Bates calls them, and again high-pitched. They resembled a bird call and in fact were imitated by a bird which lived in a nearby tree. Icty loved to dive and swim in the pool in her enclosure and retrieved sticks from the bottom of the pool with facility. She also spent a good deal of time digging and perfecting a lair in the ground inside her doghouse.

The tameability of bush dogs has been confirmed by other reports on captive animals. While it will never rival the beagle or poodle as the family pet, this adaptability to life with man raises the hope that we will eventually learn much more about the behavior of such animals from captive specimens, even if they remain as elusive as ever in the wild.

Bat-eared Fox

(Otocyon megalotis)

THE AFRICAN BAT-EARED FOX, anatomically the most peculiar member of the dog family, is sufficiently specialized to belong to its own subfamily, the *Otocyoninae.* Also called the long-eared fox, the big-eared fox, the black-eared fox and Delalande's fox, it is, of course, named for the relatively enormous ears that are its most conspicuous external feature. But other foxes have big ears; the peculiarity of this animal lies in its skeleton and teeth. It has 46 to 50 teeth, compared with 42 for most other dogs. The teeth are small and the molars relatively undifferentiated, more like those of insectivores than carnivores. The bat-eared fox is restricted to East and South Africa from southern Sudan and Ethiopia to the Cape of Good Hope.

A pretty little animal, it is 24 or 25 inches long, with a tail measuring an extra 12 or 13 inches. It stands 12 to 14 inches tall, and weighs from six to 11 lbs. Somewhat oval in shape, the ears are about four and one-half inches long—more than a third as long as the fox is tall. The animal is dark gray or brownish-gray on the back, fading to buff or whitish on the flanks, throat, and belly. The muzzle is a light brown which darkens towards the eyes; the eyelids are black, as is the chin and both fore and hind feet. The insides of the ears are white with blackish upper edges, while the backs are dark brown. The tail is buff on the lower side, but above has a brown band which broadens towards the black tip. The pupil of the eye, like that of other foxes, is a vertical slit.

The bat-eared fox has several characteristic habits, such as crouching close to the ground with its ears flat to either side

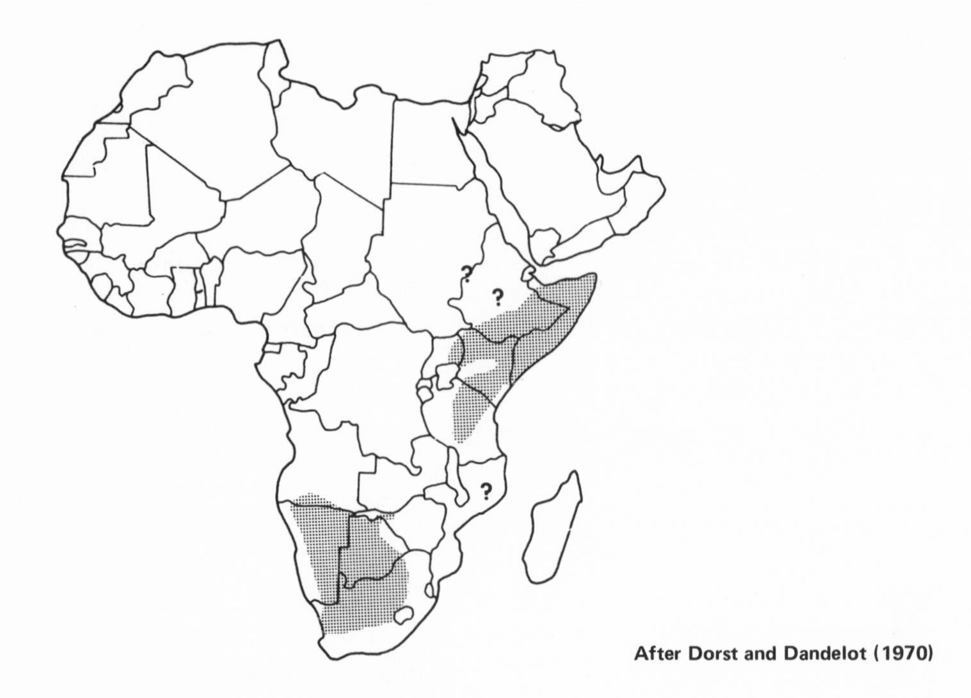

After Dorst and Dandelot (1970)

The bat-eared fox, an insect-eating native of eastern and southern Africa, typically flattens its ears when alarmed. *Photo: Sapra Studio*

when it is alarmed. It seems to pivot on one foreleg as it walks or runs, and constantly doubles back in its tracks, even when pursued, a peculiarity which helps it evade hounds, jackals, and large birds of prey. It has a slightly musky odor noticeable to people. The voice is said to be a call rather like that of the red fox, though not so loud. If cornered, it may yelp or "wuff" harshly and will growl when arguing over food. Parents reportedly call pups from the burrow with a whistling sound.

The bat-eared fox lives on the plains or in open brushland, and depends on its burrow, which it digs in open, sandy country, for protection as well as for rearing its young. Several burrows may be found close together, since this fox is a social animal which often travels about in groups of six or more and frequently raises its pups in colonies. At dawn and dusk the fox is often seen lying at the entrance of its den, and though it sometimes hunts at night is also abroad during the day.

Like so many other small dogs, the bat-eared fox eats anything it can find–rodents, birds and eggs, lizards, fruits and tubers. Unlike other dogs, however, insects form the bulk of its diet. Termites are a favorite, as are beetles, larvae, and locusts. Dung beetles, found in immense numbers where herds of grazing animals concentrate, deposit their eggs in balls of dung, several of which they bury together. The bat-eared fox seems able to hear the squeaky sound made by the larva in the dung, digs up the ball, breaks it open, and eats the grub. This fox does not often eat carrion, though it may sometimes be found poking around a deserted campsite in search of scraps.

Louis Leakey reports a remarkable feeding habit of the bat-eared fox which he first saw exhibited near the base of a cliff inhabited by a variety of small birds and by predatory kestrels and falcons. "On this particular Sunday morning, I watched a pair of lanner falcons fly along the cliff face; I saw one scare up, chase, and seize a rock pigeon–so heavy that the falcon could not carry it off. The falcon extended its wings, using them as a parachute, and started to float down with the struggling pigeon in its talons. At that point I saw something I shall never forget.

"Not far from me three bat-eared foxes were sitting outside their den. As soon as the falcon began to float down, they rushed to a point below it. As the falcon grounded, the foxes attacked it. They kept well away from its beak, but they pulled its wings and tail feathers. The falcon let go of its prey to defend itself and one of the foxes at once seized the pigeon and rushed away with it, disappearing down the den to be joined quickly by the other two.

"At the time, I considered this incident unique and accidental. Since then, I have observed the same behavior on two other occasions. I now believe that the bat-eared foxes living near cliffs make a habit of watching for occurrences of this sort and thus obtain a meal they would never be able to catch for themselves. It must make a nice change of diet from dung beetles and their larvae."

Although the bat-eared fox is a relatively well known creature, information about its reproductive habits is not complete. The gestation period seems to be about 70 days. According to some accounts, pups are born in November or December, whole others say April. The latitude can make considerable variation, of course, in the whelping season. At the London Zoo a female entered pro-estrus in mid-March and was receptive to the male for about three days in mid-April. During pro-estrus and estrus the three males and one female at the zoo increased their scent-making activities and there was increased conflict among the males. A copulatory tie was observed but there was no ensuing pregnancy. Litters of bat-eared foxes number from three to five pups, which seem to stay with the parents for several months and perhaps as adults make up part of an extended family.

The small bat-eared fox is fair game for all sorts of predators. Jackals take their toll, and large birds of prey catch pups, but the dog-loving leopard may be the greatest natural enemy. Man is an enemy too, as are his free-roving dogs. Although the bat-eared fox is a useful animal which consumes great quantities of insects and has such small teeth and weak jaws that it cannot harm livestock, too many farmers and hunters think of it as just another wild dog, bound to be vermin. It has been wiped out in many populated areas. Disease also affects the bat-eared fox population in a pattern typical of social, den-dwelling foxes. Around the Leakey camp at Olduvai Gorge in northern Tanzania there are in some years as many as 15 fox families which from time to time suddenly disappear almost completely, as though decimated by disease, possibly distemper. After such a disaster the numbers gradually re-build until a wave of disease attacks again.

Appendix I
Anatomical Charts of the Domestic Dog

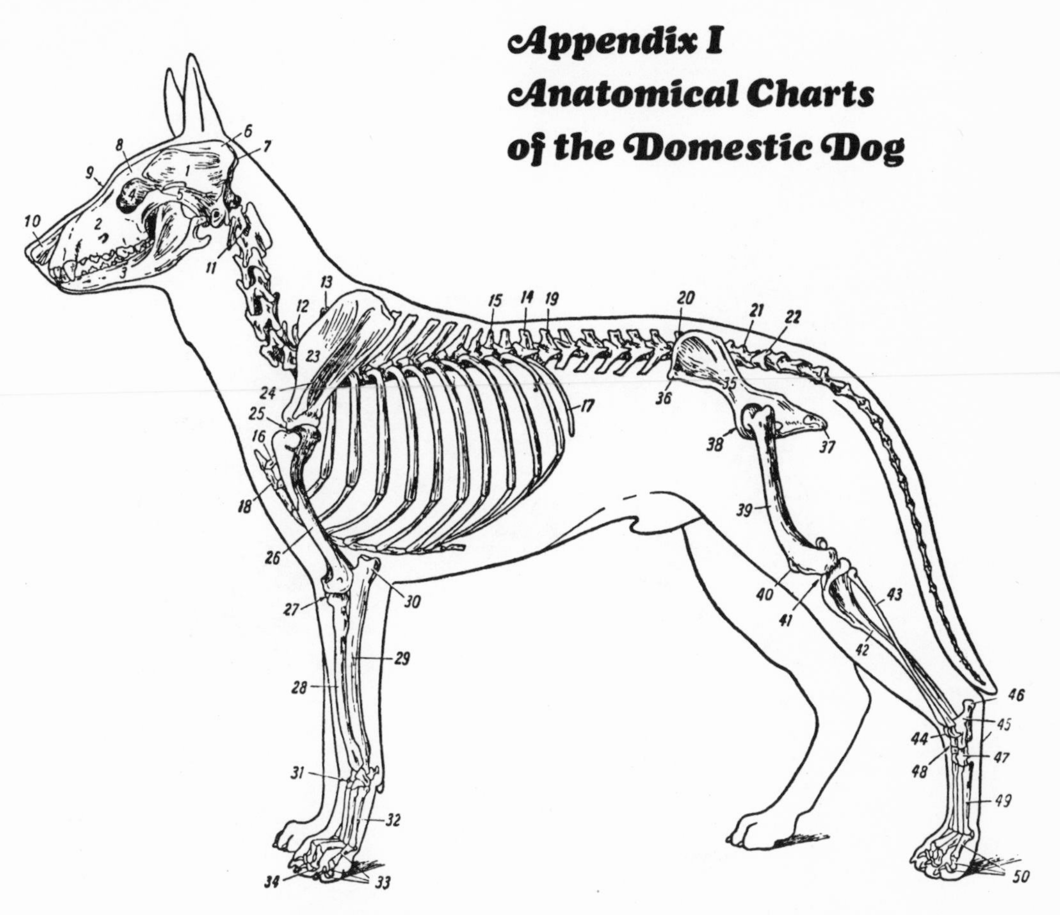

Fig. 1. THE BONE STRUCTURE OF THE DOMESTIC DOG

Bones of the Head: 1. Cranium; 2. Foreface; 3. Under jaw; 4. Eye socket; 5. Zygomatic arch; 6. Occipital crest; 7. Point of Occiput; 8. Forehead; 9. Stop; 10. Nasal cartilage.

Bones of the Trunk: 11. Spinal process of first neck vertebra; 12. Spinal process of seventh neck vertebra; 13. Spinal process of first dorsal vertebra; 14. Thirteenth dorsal vertebra; 15. Diaphragmatic vertebra; 16. First rib; 17. Thirteenth rib; 18. Sternum or breast-bone; 13-18. Chest cavity or thorax; 19. First lumbar vertebra; 20. Spine of seventh lumbar vertebra; 21. Sacrum; 22. First caudal vertebra.

Bones of the Limbs: 23. Scapula or shoulder blade; 24. Ridge on shoulder blade; 25. Shoulder joint; 26. Upper arm or humerus; 27. Elbow; 28. Radius; 29. Ulna; 30. Point of elbow; 31. Wrist (7 bones) or pastern joint; 32. Pastern (5 bones); 33. Toes of forefoot with joints; 34. Claw bone; 35. Pelvis; 36. Hip; 37. Ischium; 38. Hip joint; 39. Thigh bone or femur; 40. Knee-cap or patella; 41. Knee joint; 42. Shin bone or tibia; 43. Fibula; 44 and 48. Hock joint; 45. Heel bone; 46. Point of hock; 47. Other hock bones; 49. Middle bones of hindfoot (4 bones); 50. Toes of hindfoot with joints.

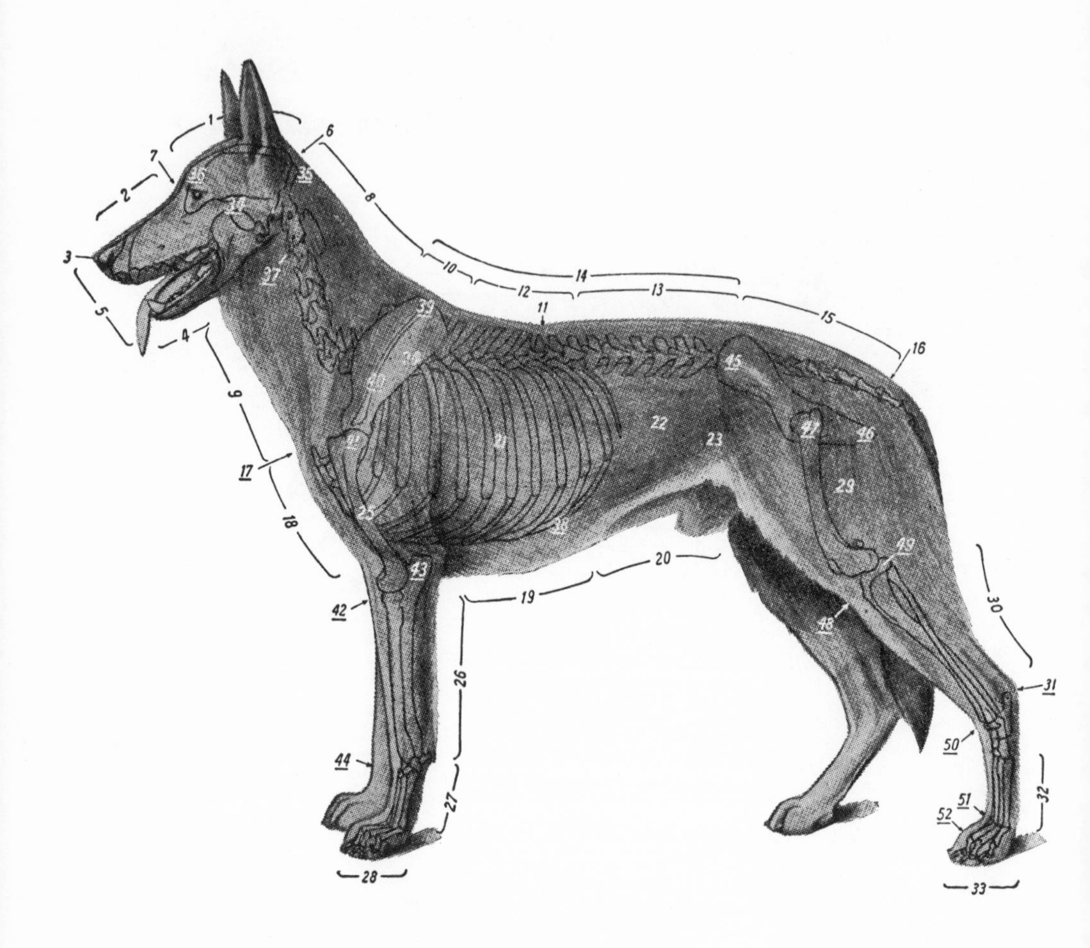

Fig. 2. THE CHIEF PARTS OF THE BODY

1. Skull; 2. Muzzle; 3. Nose; 4. Under-jaw; 5. Bite; 6. Nape; 7. Stop; 8. Nape of neck; 9. Throat; 10. Withers; 11. Small of back or back depression; 12. True back; 13. Loin; 14. Back region as a whole; 15. Croup; 16. Set-on of tail; 17. Point of breast-bone; 18. Fore chest; 19. Under chest; 20. Belly; 21. Side of chest; 22. Side of body; 23. Flank; 24. Shoulder; 25. Upper arm; 26. Lower arm or foreleg; 27. Middle portion of forefoot, often called the pastern; 28. Forefoot; 29. Thigh; 30. Second thigh; 31. Heel or point of hock; 32. Middle portion of hindfoot, loosely referred to as the hock; 33. Hindfoot; 34. Cheek; 35. Occiput; 36. Rise of skull; 37. Wing of atlas vertebra; 38. Ribs; 39. Top or point of shoulder blade; 40. Shoulder blade; 41. Shoulder joint; 42. Elbow joint; 43. Point of elbow; 44. Wrist; 45. Hip; 46. Ischium; 47. Hip joint. 48. Knee or stifle; 49. Outer projection of shin; 50. Hock joint; 51. First toe joint; 52. Second toe joint.

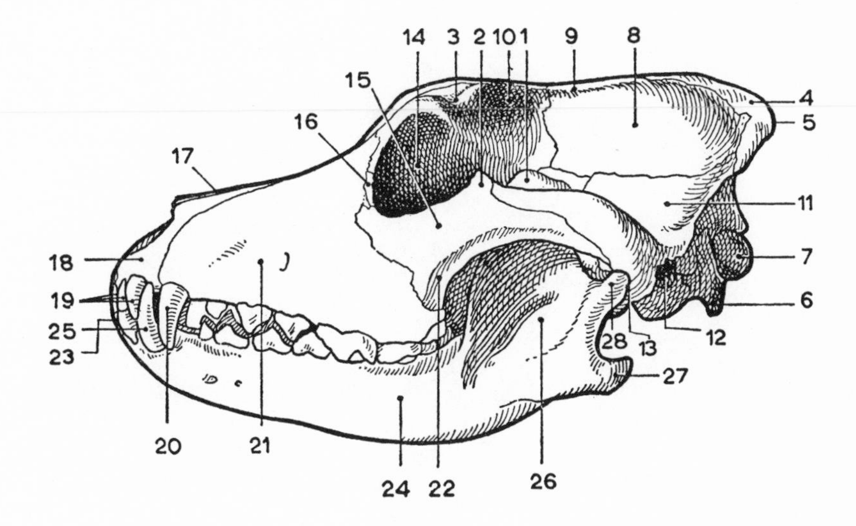

Fig. 3. THE SKULL SEEN FROM THE SIDE

1. Top of branch of lower jaw; 2. Frontal process of cheek bone; 3. Cheek process of frontal bone; 4. Occipital bone; 5. Occipital point; 6. Paroccipital process; 7. Occipital condyle; 8. Parietal; 9. Parietal or sagittal crest; 10. Frontal bone; 11. Temporal bone; 12. External auditary meatus; 13. Joint of jaw; 14. Eye socket; 15. Cheek bone or jugal; 16. Lachrymal bone; 17. Nasal bone; 18. Premaxilla; 19. Upper incisors (3 on each side); 20. Upper canine or tusk; 21. Upper jaw, each side with 6 double teeth of which the fourth and largest is the tearing or rending tooth (carnassial); 22. Ridge on cheek bone; 23. Unpaired part of under-jaw with 6 lower incisors; 24. Paired part of under-jaw with 7 lower double teeth on each side, of which the fifth and largest is the tearing tooth; 25. Lower canine or tusk; 26. Ramus of lower jaw; 27. Hind process of lower jaw; 28. Articulating process of lower jaw.

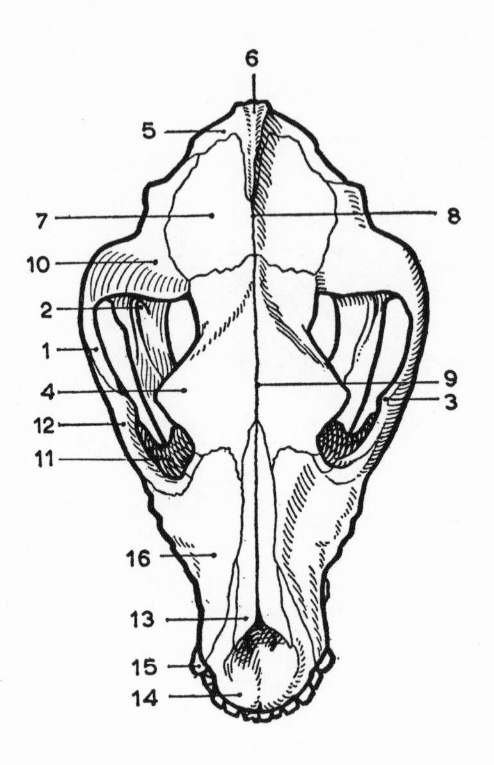

Fig. 4. THE SKULL SEEN FROM ABOVE

1. Zygomatic arch, which is especially prominent in short-headed breeds. The stronger the curve of this arch the greater the width of the skull; 2. Ramus of lower jaw; 3. Frontal process of cheek bone; 4. Jugal projection on frontal; 5. Occipital; 6. Occipital point, which is more or less visible, or can be felt, in all breeds; 7. Parietal bone; 8. Parietal or sagittal ridge or crest, usually somewhat prominent in long-headed dogs but wanting in short-headed breeds. Discernible as a slight ridge among the strong muscles of the cranium; 9. Frontal bone with lateral jugal process; 10. Temporal bone; 11. Eye socket or orbit; 12. Cheek bone or jugal; 13. Nasal bones, the foundation of the muzzle; also connect with the nose; 14. Premaxilla with tooth sockets for the upper incisors; 15. Upper canine or tusk; 16. Upper jaw bone or maxilla.

Left, from below *Right*, from the side

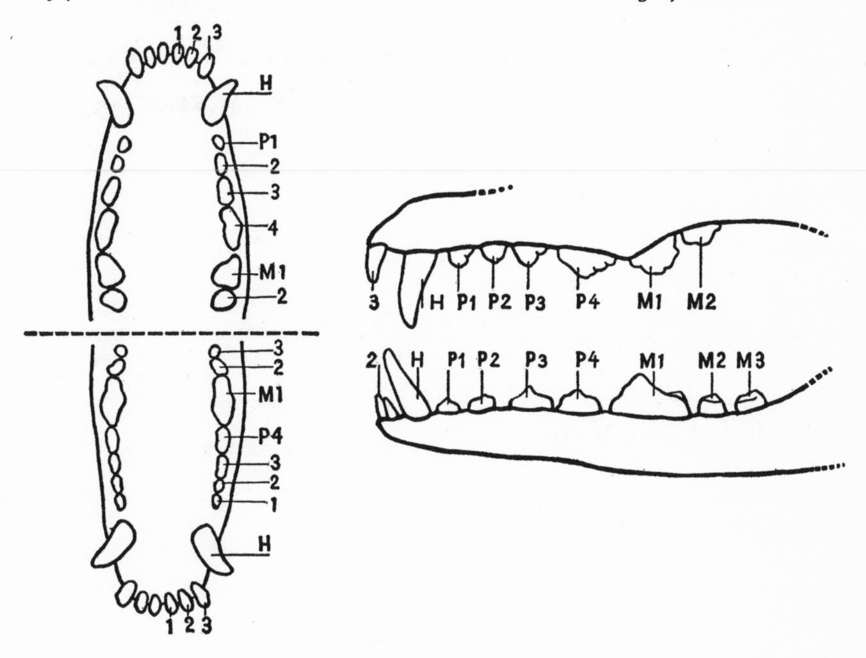

Fig. 5. THE BITE

The complete bit comprises 42 teeth, 20 in the upper jaw and 22 in the lower jaw. Each side of the jaw includes:

In the upper jaw	In the lower jaw	
3	3	Cutting or incisor teeth, 1, 2, 3
1	1	Canine tooth or tusk H
4	4	Premolar teeth, P1 to P4
2	3	Molar teeth, M1 and 2 and M1, 2, 3

The fourth premolar in the upper jaw and the first molar in the lower jaw are tearing teeth or carnassials. The cutting teeth are used mainly for gnawing bones and for attending to the coat. The canines grip and hold prey, the tearing teeth and premolars reduce the food to small pieces and the molars chew.

Appendix II
Scientific Names and Taxonomic Status

A—COMMON AND SCIENTIFIC NAMES OF ALL SPECIES OF DOGS

Gray Wolf	*Canis lupus*
Red Wolf	*Canis rufus*
Coyote	*Canis latrans*
Golden Jackal	*Canis aureus*
Black-backed Jackal	*Canis mesomelas*
Side-striped Jackal	*Canis adustus*
Dingo	*Canis dingo*
Domestic Dog	*Canis familiaris*
Semyen Fox	*Simenia simensis*
Red Fox	*Vulpes vulpes*
Kit Fox	*Vulpes velox*
Corsac Fox	*Vulpes corsac*
Bengal Fox	*Vulpes bengalensis*
Hoary Fox	*Vulpes cana*
Tibetan Sand Fox	*Vulpes ferrilata*
Rüppell's Sand Fox	*Vulpes rüppelli*
African Sand Fox	*Vulpes pallida*
Cape Fox	*Vulpes chama*
Arctic Fox	*Alopex lagopus*
Fennec	*Fennecus zerda*
Gray Fox	*Urocyon cinereoargenteus*
Culpeo	*Dusicyon culpaeus*
Santa Elena Fox	*Dusicyon culpaeolus*
Pampas Fox	*Dusicyon gymnocercus*

Peruvian Fox	*Dusicyon inca*
Chilla	*Dusicyon griseus*
Chiloé Fox	*Dusicyon fulvipes*
Field Fox	*Dusicyon vetulus*
Sechura Fox	*Dusicyon sechurae*
Forest Fox	*Cerdocyon thous*
Small-eared Dog	*Atelocynus microtis*
Maned Wolf	*Chrysocyon brachyurus*
Raccoon Dog	*Nyctereutes procyonoides*
African Hunting Dog	*Lycaon pictus*
Dhole	*Cuon alpinus*
Bush Dog	*Speothos venaticus*
Bat-eared Fox	*Otocyon megalotis*

B–TAXONOMIC STATUS AND NOMENCLATURE OF SOME SPECIES

The taxonomic status of several species of wild dogs is rather confused. In a few cases species have undergone a recent change of scientific name. I have listed by species those animals for which a clarification of nomenclature and current status might be useful.

Red Wolf (*Canis rufus*)
Until recently the red wolf was called *Canis niger. Niger* had been the species name used by Young and Goldman (1944) based on its use by William Bartram in his *Travels* (1791), a description of the southeastern United States. However, Nowak (1967) pointed out that according to the International Commission on Zoological Nomenclature the names used by Bartram are not available, and that the specific name should be *rufus. Canis rufus* is now accepted as the correct scientific name and is used in most recent literature.

The Jackals (*Canis aureus, C. mesomelas, C. adustus*)
The three subspecies of jackals have sometimes been given the generic or subgeneric name *Thos.* The Egyptian jackal, occasionally encountered under the name *C. lupaster,* should be considered a race of *C. aureus.*

Dingo (*Canis dingo*)
The dingo, a feral dog, is sometimes grouped with the domestic dog under the name *C. familiaris dingo.* The New Guinea singing dog, described as a separate species *C. hallstromi,* should be considered a dingo or feral dog.

Domestic Dog (*Canis familiaris*)
Living domestic dogs are not divided into subspecies. However, zoologists and archeologists have sometimes assigned subspecific names to remains of Stone or Bronze Age animals. While some of these supposed races have

absolutely no basis in fact, several of the better-known designations correspond to recurring size or skeletal characteristics and are frequently encountered in works on domestic dogs.

The so-called *Canis familiaris poutiatini* refers to Neolithic animals of medium or dingo size, the type specimen of which was found in northern Russia; these animals are supposed to have given rise to two other types, *C. f. intermedius* and *C. f. matris-optimae. Matris-optimae,* a Bronze Age dog, is supposed to be the forerunner of the sheep and herd dogs of Europe, while *intermedius,* also dating from the Bronze Age, has been regarded by older writers as the ancestor of the hounds.

Another technical name often encountered is *C. f. inostranzewi.* This type, originally unearthed at Lake Ladoga in Russia, was a large animal supposed to represent a cross between *poutiatini* and northern wolves, and which has been called the ancestral type of present Nordic breeds. Finally, the name *C. f. palustris,* the Peat or Lake Dog of the Neolithic lake dwellers, is frequently listed in the literature. This *palustris* is perhaps the most distinct type here mentioned; the name refers to a smaller dog than others of Neolithic Europe and many writers have seen in it the ancestor of the house dogs and toy breeds.

Modern zoologists find much fault with this assigning of subspecific names to often incomplete or questionable remains, and the supposed descent of modern breeds can likewise often be called into question. For instance, *C. f. matris-optimae,* the Bronze Age dog which has been called the forerunner of the hounds, postdates the historical representations and skeletal remains of the coursing hounds of the Near East and Egypt. The evidence which links *inostranzewi* with the wolves of northern Europe is extremely conjectural and an unnecessary explanation of the size of northern breeds.

Semyen Fox (*Simenia simensis*)

There is still a question about whether the Semyen fox should be considered part of the genus *Canis* or should be accorded separate generic status as *Simenia.* Current work being carried out by various researchers may soon clarify the animal's status. Until then I have chosen to signal its anomolous position by the use of the name *Simenia.*

Red Fox (*Vulpes vulpes*)

The North American red fox, formerly called *Vulpes fulva,* is now considered the same species as the Old World red fox *V. vulpes.* Basic information on the relationship between Old World and New World red foxes can be found in Churcher (1959) and Gilmore (1946), and a more recent summary in Burrows (1968).

Kit Fox (*Vulpes velox*)

The desert race of the kit fox, sometimes called the swift fox, has frequently been considered a separate species under the name *V. macrotis*. Now it is generally agreed that there are no grounds for giving these two races more than subspecific status.

South American Foxes (*Dusicyon* and *Cerdocyon*)

The naming of the South American foxes is still chaotic and taxonomic information frequently contradictory. I have generally followed the work of Cabrera (1957; Cabrera and Yepes 1940) because his is the most recent attempt to wrestle with the classification of South American dogs as a whole. It should be noted that Cabrera regards *Pseudalopex* Burmeister as a synonym for *Dusicyon*, and *Lycalopex* Burmeister as a subgeneric name for the field fox (*D. vetulus*) and the Sechura fox (*D. sechurae*).

Small-eared Dog (*Atelocynus microtis*)

The small-eared dog clearly is entitled to its status as a separate genus, and the scientific name *Atelocynus microtis* seems now to be generally accepted. It is helpful to know that this name is synonymous with *Dusicyon microtis* Osgood and *Cerdocyon microtis* Hoffstetter.

Maned Wolf (*Chrysocyon brachyurus*)

Older sources sometimes use the specific name *jubatus* for the maned wolf; however, *Chrysocyon brachyurus* is currently considered the proper designation.

Raccoon Dog (*Nyctereutes procyonoides*)

Like the maned wolf, the raccoon dog is difficult to confuse with any other animal. However, the Japanese subspecies has in older sources often been described under the name *Nyctereutes viverrinus*. In keeping with current taxonomic trends, *N. p. viverrinus* is now considered only a race of the raccoon dog.

Dhole (*Cuon alpinus*)

Present opinion holds that the genus *Cuon* (sometimes called *Cyon*) includes only one species, designated *alpinus*, though in the past the more southern races have been grouped in one or several species referred to as *primoevus, sumatrensis, dukhunensis, rutilans* or *javanicus*. From the modern point of view the differences among the various races are not sufficient to constitute separate species.

Bush Dog (*Speothos venaticus*)

The northwestern subspecies of the bush dog, originally identified from Pirré hill in the extreme eastern part of Panama, was initially designated a separate species, *Speothos panamensis*. It is now regarded as only a subspecies of the bush dog under the name *S. v. panamensis*. The generic name *Speothos* replaces the older designation *Icticyon*.

Appendix III
Relationships Among Members of the Dog Family

The family of dogs, or Canidae, has traditionally been divided into three subfamilies: the Caninae, including most species; the Simocyoninae, including *Lycaon* (African hunting dog), *Cuon* (dhole) and *Speothos* (bush dog); and the Otocyoninae, comprised of the single species *Otocyon megalotis,* the bat-eared fox.

The bat-eared fox has been placed apart because of its dental peculiarities: four to eight more teeth than any other dog, and so relatively unspecialized as to be more like the teeth of Insectivora than of other dogs.

The three species of the Simocyoninae also have dental peculiarities which account for their distinctive grouping: the first molar on the lower jaw is different from that of the Caninae. In the latter this molar, in addition to being relatively large, has a basin-like depression at the posterior end bounded by two distinct cusps. In the African hunting dog, the dhole, and the bush dog, this tooth is relatively smaller and lacks the basin-like depression; on its heel the inner cusp is reduced to near non-existence. This molar is thus described as trenchant in the Simocyoninae, while in the Caninae it is referred to as basined. In the dhole and the bush dog the number of teeth is also reduced—the dhole lacks the third lower molar (M_3) and has 40 teeth in all, while the bush dog lacks both the third lower and the second upper molars (M_3 and M^2) for a total of 38 teeth.

The trenchant character of the heel of the first lower molar is a very old one in dogs, dating back to the Oligocene. Matthew (1930), who did extensive work on the phylogeny of dogs, considered that modern dogs having the characteristic were the survivors of an old group once widespread throughout the world which had gradually been pushed south by a more recent and northern group, the Caninae. Matthew's conclusions have generally been followed by other taxonomists.

A problem with the Simocyoninae is that the three wild dogs thus grouped have nothing specific in common except this one tooth characteristic. Though two of the three species show a reduction in the number of teeth, the African hunting dog does not. The latter has a peculiarity—the absence of the pollex or dew-claw—not shared by the other two. While the dhole and the African hunting dog are relatively large, long-legged animals which pursue their prey at speed, the bush dog is small, stubby-legged, and heavy, evidently as dependent on swimming as running to capture prey. Also, the bush dog has a short, straight caecum or blind intestine like that of two other South American dogs, rather than the long, coiled caecum common to most canids—including the hunting dog and the dhole.

Since the work of Matthew classification of the Simocyoninae has been questioned by several workers, including Pocock (1941), Hough (1948), and Thenius (1954). Hough noted that the trenchant heel of the lower carnassial is a characteristic which has independently appeared several times in carnivores and need not indicate a genetic relationship. Thenius examined several recent fossil forms of *Cuon* and showed that in this genus the trenchant heel of the one molar and the loss of the other are recent developments, occurring during the middle and late Quarternary. Both Thenius and Hough suggested that on the basis of physical characteristics and behavior the dhole resembles the jackals, the African hunting dog the wolves, and the bush dog the South American canids.

Much work needs to be done on relationships among living dogs. I have retained the traditional groupings to the extent of placing the chapters on the African hunting dog, the dhole, and the bush dog after those dealing with the other members of the family, and have concluded the body of the book with the description of the bat-eared fox, the most clearly aberrant dog.

Within the Caninae, I have grouped the various species by genera, and in cases in which a genus includes only a single species, have placed it adjacent to other animals with which it has the most in common. Thus the chapter on the arctic fox follows those on the genus *Vulpes,* for anatomically the arctic fox is very similar to the vulpine foxes. With the exception of the bush dog, the various South American dogs are grouped together and the chapter on the gray fox (*Urocyon*) immediately precedes them, since unquestionably the gray fox has more in common with the South American genus *Dusicyon* than it does with dogs of the Old World. In only one case have I included a description of an extinct animal. A few lines are devoted to the Falkland fox (*D. australis*) in the section on the genus *Dusicyon.* The Falkland fox became extinct a little less than a hundred years ago, as a direct result of hunting by man.

Appendix IV
Genetic Information

A—CHROMOSOME NUMBERS OF SOME SPECIES

Only a few studies of chromosome numbers and characteristics have been made for members of the dog family, and their results are sometimes contradictory. The following list gives diploid chromosome numbers and sources for seven species.

Golden Jackal (*Canis aureus*)—74 (Matthey, 1954)
Domestic Dog (*Canis familiaris*)—78 (Matthey, 1954; Ahmed, 1941, cited in Scott and Fuller, 1965; among others)
Red Fox (*Vulpes vulpes*)—34 (Wipf and Shackelford, 1949); older studies have given other numbers—Wadsedalek, 42; Bishop, 32; Makino, 38 (cited in Wipf and Shackelford)
Rüppell's Sand Fox (*Vulpes rüppelli*)—40 (Matthey, 1954)
Arctic Fox (*Alopex lagopus*)—52 (Wipf and Shackelford, 1949)
Fennec (*Fennecus zerda*)—64 (Matthey, 1954)
Raccoon Dog (*Nyctereutes procyonoides*)—42 (Minouchi, 1929)

B—INTER-SPECIFIC CROSSES AMONG DOGS

The ability of gray wolves and coyotes to interbreed with domestic dogs to produce fertile hybrids is well established. The possibility of fertile crossings between domestic dogs and the several species of jackals has been a matter of dispute for decades and available evidence is still contradictory. Reports of inter-specific crosses among other wild dogs occur occasionally but are for the most part not scientifically confirmable. All claims of interspecific crosses and of the fertility of hybrids must be regarded with scepticism unless crosses have been made repeatedly under scientifically controlled conditions.

The following list of inter-specific crosses differentiates among cases which are well confirmed, those which rest on one or a few scientifically reported crosses, and those which are unconfirmed.

Gray Wolf (*Canis lupus*)

The wolf crosses readily with the domestic dog and the hybrids are fertile. Crosses with various proportions of wolf and dog blood have been carried to many generations. It is assumed that wolves would interbreed readily with both coyotes and red wolves to produce fertile offspring.

Red Wolf (*Canis rufus*)

Red wolves have hybridized extensively with coyotes in the western part of their original range.

Coyote (*Canis latrans*)

Coyotes cross readily with the domestic dog, and the hybrids, called coy-dogs, are fertile. Coyotes have hybridized with the red wolf in the wild, and it may be assumed that they would interbreed fertilely with the gray wolf. Herre (1964) describes well-documented crosses between the coyote and the golden jackal which were carried to the third generation.

Golden Jackal (*Canis aureus*)

Reports of golden jackal-domestic dog crosses have appeared from time to time, the most interesting of the older ones being that of Kühn (1887), who reported a cross in which the hybrids were fertile with each other and when backcrossed to the dog (cited in Allen, 1920); and that of Hilzheimer, who reputedly produced a triple hybrid of wolf-jackal-domestic dog (cited in Iljin, 1941, from Lotsy, 1922). Such older reports were called into question by the work of Matthey (1954), which indicated that while the domestic dog has 78 chromosomes the golden jackal has only 74. On the other hand, the experiment described by Herre (1964) which produced several generations from coyote-golden jackal crosses makes the likelihood of fertile jackal-dog crosses greater. Obviously there is a need for hybridization experiments between dogs and jackals to be carried out under carefully controlled conditions, and Matthey's work on chromosome numbers also needs to be confirmed before the genetic relationship between the two species can be clarified.

Black-backed Jackal (*Canis mesomelas*)

Van der Merwe (1954) describes several cases of domestic dog-black-backed jackal crosses in captivity reported to him by correspondents. None of these cases was independently confirmed, nor was any information available on the fertility of the hybrids.

Red Fox (*Vulpes vulpes*)

Crosses have been made by fur breeders between the red and the arctic fox but no fertile hybrids have been obtained. Wipf and Shackelford (1949) repeated the cross under controlled conditions and examined the chromosomes of the parent species and the hybrid. They found that the red fox has 34 chromosomes, the arctic fox 52, and the hybrid had 43. The sterility of the hybrid resulted from the inability of the chromosomes to pair properly to produce viable gametes.

Arctic Fox (*Alopex lagopus*)

See Red Fox.

Pampas Fox (*Dusicyon gymnocercus*)

Krieg (1925; cited in Iljin, 1941) described two litters which were supposed to have resulted from a cross between a male fox terrier and a female South American fox which he called the pampas fox *Pseudalopex* (*Canis*) *azarae.* Although the nomenclature of the various South American foxes is very confused, he may have been referring to the pampas fox *Dusicyon gymnocercus.* At any rate this reputed cross has not been repeated by other researchers.

Forest Fox (*Cerdocyon thous*)

It has been rumored for at least 100 years that the forest fox can be crossed with the domestic dog. Darwin reported that the domestic dogs of Guiana could be crossed with *Cerdocyon,* called locally the *maikong* and for which he used the scientific name *Canis cancrivorus.* Cabrera (1940) refers to the fact that the forest fox is quite tameable and that Indians of the Guianas are said to obtain useful hunting dogs by crossing their domestic dogs with this fox. Finally, Walsh (1967) describes a little dog he adopted in Surinam which was reported by the natives to be the result of a mating between a female domestic dog and some sort of wild dog, perhaps the forest fox. Whatever the origin of Walsh's dog, his account indicates that the rumor is still current.

Bibliography

Allen, Glover M. "Dogs of the American Aborigines." *Bulletin of the Museum of Comparative Zoology* (Harvard College), 63:429–517 (1920).

Asdell, S. A. *Patterns of Mammalian Reproduction.* 2d ed. Ithaca, N.Y.: Cornell University Press, 1964.

Bates, Marston. *The Land and Wildlife of South America.* Life Nature Library. New York: Time Inc., 1964.

_____. "Notes on a Captive *Ictycyon.*" *Journal of Mammalogy* 25:152–54 (1944).

Bourlière, François. *Mammals of the World, Their Life and Habits.* New York: Alfred A. Knopf, 1955.

_____. "Systématique," in *Traité de Zoologie,* Vol. XVII, *Mammifères.* Paris: Masson et Cie., 1955.

_____. *The Land and Wildlife of Eurasia.* Life Nature Library. New York: Time Inc., 1964.

Brown, Leslie. *Ethiopian Episode.* London: Country Life Ltd., 1965.

*Burrows, Roger. *Wild Fox.* Newton Abbot, England: David and Charles, 1968.

Cabrera, Angel. "Catalogo de los Mamiferos de America del Sur. "*Revista del Museo Argentino de Ciencias Naturales.*" *ciencia Zoológicas,* Tomo IV (1), 1957. 307 pp.

_____, and José Yepes. *Mamiferos Sud-americanos.* Buenos Aires: Compania Argentina de Editories, 1940.

Cade, C. E. "Notes on Breeding the Cape Hunting Dog *Lycaon pictus* at Nairobi Zoo." *International Zoo Yearbook* 7:122–23 (1967).

Churcher, C. S. "The Specific Status of the New World Red Foxes." *Journal of Mammalogy* 40:513–20 (1959).

Coimbra Filho, A. F. "Notes on the Reproduction and Diet of Azara's Fox *Cerdocyon thous azarae* and the Hoary Fox *Dusicyon vetulus* at Rio de Janeiro Zoo." *International Zoo Yearbook* 6:168–69 (1966).

*Works marked with an asterisk are particularly interesting additional reading.

Cole, Glen F. "Gray Wolf Population Status and Trends." Research note. Yellowstone National Park Memorandum. Washington, D.C.: U.S. Dept. of Interior, National Park Service, 1969. 5 pp.

*Crisler, Lois. *Arctic Wild.* New York: Harper and Row, 1958.

Cutter, William L. "Denning of the Swift Fox in Northern Texas." *Journal of Mammalogy* 39:70-74 (1958).

Darlington, P. J. *Zoogeography.* New York: John Wiley & Sons, 1957.

Dathe, Heinrich. "Breeding the Corsac Fox *Vulpes corsac* at East Berlin Zoo." *International Zoo Yearbook* 6:166-67 (1966).

Dekker, D. "Breeding the Cape Hunting Dog *Lycaon pictus* at Amsterdam Zoo." *International Zoo Yearbook* 8:27-30 (1968).

*Dobie, J. Frank. *The Voice of the Coyote.* Omaha: University of Nebraska Press, 1961. (Originally published by Little, Brown and Co., Boston, 1949.)

Dorst, Jean, and Pierre Dandelot. *Field Guide to the Larger Mammals of Africa.* Boston: Houghton Mifflin, 1970.

Eaton, Randall L. "Cooperative Hunting by Cheetahs and Jackals and a Theory of Domestication of the Dog." *Mammalia* 33:87-92 (1969).

Egoscue, Harold J. "Preliminary Studies of the Kit Fox in Utah." *Journal of Mammalogy* 37:351-57 (1956).

Ellerman, J. R., and T. C. S. Morrison-Scott. *Checklist of Palaearctic and Indian Mammals.* London: British Museum, 1951.

Fentress, John C. "Observations on the Behavioral Development of a Hand-reared Male Timber Wolf." *American Zoolologist* 7:339-51 (1967).

Fisher, Arthur H. "The Fennec." *Nature Magazine* 33:576 (1940).

*Fox, Michael W. *Behaviour of Wolves, Dogs and Related Canids.* London: Jonathan Cape, 1971.

Freuchen, Peter, and Finn Salomonsen. *The Arctic Year.* New York: G. P. Putnam's Sons, 1958.

Gangloff, L. "Breeding Fennec Foxes *Fennecus zerda* at Strasburg Zoo." *International Zoo Yearbook* 12:115-16 (1972).

Geptner, Vladimir Georgievich, and Nikolai Pavlovich Naumov, eds. *Mlekoptaiuschie Sovetskogo Soiuza* (*Mammals of the Soviet Union*). 3 vols. Moscow: Gosudarstvennoe izdatelstvo "Vysshaia shkola," 1967.

Gier, H. T., and D. J. Ameel. "Parasites and Diseases of Kansas Coyotes." *Kansas State University Agricultural Experiment Station Technical Bulletin* 91, 1959. 34 pp.

Gilfillian, Archer B. *Sheep.* Boston: Little, Brown and Co., 1929. (Cited in Dobie, 1961.)

Gilmore, R. M. "Mammals in Archeological Collections from S.W. Pennsylvania." *Journal of Mammalogy* 27:227-35 (1946).

Goodwin, George C. *Mammals,* Vol. I, *The Animal Kingdom.* Garden City, N.Y.: Doubleday and Co., 1954.

Guggisberg, C. A. W. "Wild Dogs of the Savannah." *Animals* 1(24):4-7 (1963).

Haag, William G. "An Osteometric Analysis of Some Aboriginal Dogs." *University of Kentucky Reports in Anthropology* 7(3), 1948. 264 pp.

Hall, E. Raymond, and Keith R. Kelson. *The Mammals of North America.* 2 vols. New York: The Ronald Press Co., 1959.

Harrison, David L. *The Mammals of Arabia,* Vol. II. London: Ernest Benn Ltd., 1968.

Herre, Wolf. "Demonstration im Tiergarten des Institutes für Haustierkunde der Universität Kiel, insbesondere von Wildcaniden und Caniden-Kreuzungen (Schakal/Coyoten F_1- und F_2-Bastarde sowie Pudel/Wolf-Kreuzungen)." *Zoologischer Anzeiger* 28, Supplement: 622-35 (1964).

Hershkovitz, Philip. "On the South American Small-eared Zorro *Atelocynus microtis* Sclater (Canidae)." *Fieldiana: Zoology* (Field Museum of Natural History), 39:505-23 (1961).

———. "A Synopsis of the Wild Dogs of Colombia." *Novedades Colombianas,* Museo de Historia Natural de la Universidad del Cauca 3:157-61 (1958).

Hildebrand, Milton. "Comparative Morphology of the Body Skeleton in Recent Canidae." *University of California Publications in Zoology* 52:399-470 (1954).

Hough, J. R. "The Auditory Region in Some Members of the Procyonidae, Canidae, and Ursidae." *Bulletin of the American Museum of Natural History* 92:67-118 (1948).

Iljin, N. A. "Wolf-Dog Genetics." *Journal of Genetics* 42:359-414 (1941).

Jolicoeur, Pierre. "Multivariate Geographical Variation in the Wolf *Canis lupus* L." *Evolution* 13:283-99 (1959).

Jordan, Peter A., Philip C. Shelton, and Durwood L. Allen. "Numbers, Turnover, and Social Structure of the Isle Royale Wolf Population." *American Zoologist* 7:233-52 (1967).

Joslin, Paul W. B. "Movements and Home Sites of Timber Wolves in Algonquin Park." *American Zoologist* 7:279-88 (1967).

Kitchener, Saul L. "Observations on the Breeding of the Bush Dog *Speothos venaticus* at Lincoln Park Zoo, Chicago." *International Zoo Yearbook* 11:99-101 (1971).

Kleiman, Devra. "Reproduction in the Canidae." *International Zoo Yearbook* 8:3-8 (1968).

———. "Some Aspects of Social Behavior in the Canidae." *American Zoologist* 7:365-72 (1967).

Kolenosky, George B., and David H. Johnston. "Radio-Tracking Timber Wolves in Ontario." *American Zoologist* 7:289-303 (1967).

Krieg, H. "Biologische Reisestudien in Sudamerika. 7. Not. über einen Bastard zwischen Hund und Pampafuchs (*Pseudalopex* [*Canis*] *azarae*) nebst einige Bemerkungen über Systematik der argentinisch-chilenischen Füchse." *Zeitschrift für Morphologie und Ökologie der Tiere* 4:702-10 (1925). (Cited in Iljin, 1941.)

Krieg, Hans. "Like a Fox on Stilts." *Animals* 3:599-601 (1964).

Kruuk, Hans, and M. Turner. "Comparative Notes on Predation by Lion, Leopard, Cheetah and Wild Dog in the Serengeti Area, East Africa." *Mammalia* 31:1-27 (1967).

Kühme, Wolfdietrich. "Communal Food Distribution and Division of Labour in African Hunting Dogs." *Nature* 205:443-44 (1965).

Lawrence, Barbara, and William H. Bossert. "Multiple Character Analysis of *Canis lupus, latrans* and *familiaris,* with a Discussion of the Relationships of *Canis niger." American Zoologist* 7:223-32 (1967).

Leakey, Louis S. B. *The Wild Realm: Animals of East Africa.* Washington, D.C.: National Geographic Society, 1969.

Longman, Heber A. "Notes on the Dingo, the Wild Indian Dog, and a Papuan Dog." *Memoirs of the Queensland Museum* 9:151-57 (1928).

Lorenz, Konrad. *Man Meets Dog.* Harmondsworth, England: Penguin Books Ltd., 1964. (First published in Austria in 1953 under the title *So Kam der Mensch auf den Hund.*)

_____. "What Aggression Is Good For: Part I." *Animals* 9:339-43 (1966).

Maberly, C. T. Astley. *Animals of East Africa.* Cape Town, South Africa: Howard Timmins, 1962.

McCarley, Howard. "The Taxonomic Status of Wild Canis (Canidae) in the South Central United States." *The Southwestern Naturalist* 7:227-35 (1962).

McIntosh, D. L. "Food of the Fox in the Canberra District." CSIRO *Wildlife Research* 8:1-20 (1963).

_____. "Reproduction and Growth of the Fox in the Canberra District." CSIRO *Wildlife Research* 8:132-41 (1963).

Matera, E. A., A. M. Saliba, and A. Matera. "The Occurrence of Dioctophymiasis in the Maned Wolf." *International Zoo Yearbook* 8:24-27 (1968).

Matthew, W. D. "The Phylogeny of Dogs." *Journal of Mammalogy* 11:117-38 (1930).

Matthey, Robert. "Chromosomes et systématique des Canidés." *Mammalia* 18:225-30 (1954).

*Mech, L. David. *The Wolf: The Ecology and Behavior of an Endangered Species.* Garden City, N.Y.: The Natural History Press, 1970.

_____. "The Wolves of Isle Royale." *Fauna of the National Parks of the U.S., Fauna Series #7.* Washington, D.C.: U.S. Govt. Printing Office, 1966. 210 pp.

Merwe, N. J. van der. "The Jackal." *Fauna and Flora Series* #4. Pretoria, South Africa: Transvaal Provincial Administration, 1953.

Minouchi, O. "On the Spermatogenesis of the Racoon Dog (*Nyctereutes viverrinus*), with Special Reference to the Sex-Chromosomes." *Cytologia* 1:88-108 (1929).

Mivart, St. George. *Dogs, Jackals, Wolves and Foxes. A Monograph of the Canidae.* London, 1890. 216 pp.

*Murie, Adolph. *A Naturalist in Alaska.* New York: Devin-Adair Co., 1961.

Novikov, G. A. "Carnivorous Mammals of the Fauna of the USSR." *Keys to the fauna of the USSR* published by the Zoological Institute of the Academy of Sciences of the USSR, Moscow, #62, 1956. (Translated by Israel Program for Scientific Translations, Jerusalem, 1962.)

Nowak, R. M. "The Mysterious Wolf of the South." *Natural History* 81:50-53; 74-77 (1972).

_____. "The Red Wolf in Louisiana." *Defenders of Wildlife News* 42:60-70 (1967).

Ogilvie, Philip W. "Interim Report on the Red Wolf in the United States." *International Zoo Yearbook* 10:122-24 (1970).

Paradiso, John L. "Canids Recently Collected in East Texas, with Comments on the Taxonomy of the Red Wolf." *The American Midland Naturalist* 80:529-34 (1968).

Pimlott, Douglas H. "Wolf Predation and Ungulate Populations." *American Zoologist* 7:267-78 (1967).

_____, and Paul W. Joslin. 1968. "The Status and Distribution of the Red Wolf." *Transactions of the 33rd North American Wildlife and Natural Resources Conference,* pp. 373-89. Published by the Wildlife Management Institute, Washington, D.C.

Pocock, R. I. *The Fauna of British India,* Vol. II, *Mammalia.* London: Taylor and Francis, Ltd., 1941.

Pulliainen, Erkki. "A Contribution to the Study of the Social Behavior of the Wolf." *American Zoologist* 7:313-17 (1967).

_____. "Studies on the Wolf (*Canis lupus* L.) in Finland." *Annales Zoologici Fennici* 2:215-59 (1965).

Rabb, George B., Jerome H. Woolpy, and Benson E. Ginsburg. "Social Relationships in a Group of Captive Wolves." *American Zoologist* 7:305-11 (1967).

Rausch, Robert A. "Some Aspects of the Population Ecology of Wolves, Alaska." *American Zoologist* 7:253-65 (1967).

Rosenberg, Hans. "Breeding the Bat-eared Fox *Otocyon megalotis* at Utica Zoo." *International Zoo Yearbook* 11:101-02 (1971).

*Rue, Leonard Lee, III. *The World of the Red Fox.* Living World Books. Philadelphia: J. B. Lippincott Co., 1969.

*Rutter, Russell J., and Douglas H. Pimlott. *The World of the Wolf.* Living World Books. Philadelphia: J. B. Lippincott Co., 1967.

Schenkel, Rudolph. "Submission: Its Features and Function in the Wolf and Dog." *American Zoologist* 7:319-29 (1967).

Schneider-Leyer, Erich. *Dogs of the World.* Translated and with five additional chapters on management by E. Fitch Daglish. London: Popular Dogs, 1964. (Originally published under the title *Die Hunde der Welt* by Albert Müller Verlag, Zurich, 1960.)

*Scott, John Paul, and John L. Fuller. *Genetics and the Social Behavior of the Dog.* Chicago: University of Chicago Press, 1965.

Silveira, Estanislau K. P. da. "Notes on the Care and Breeding of the Maned Wolf *Chrysocyon brachyurus* at Brasilia Zoo." *International Zoo Yearbook* 8:21-23 (1968).

Smithsonian Institution. *Preliminary Identification Manual for African Mammals.* J. Meester, ed. Chap. 7, "Carnivora (excluding Felidae)," by C. G. Coetzee. Washington, D.C., 1967.

Snow, Carol J. "Some Observations on the Behavioral and Morphological Development of Coyote Pups." *American Zoologist* 7:353–55 (1967).

Sosnovskii, Igor P. "Breeding the Red Dog or Dhole *Cuon alpinus* at Moscow Zoo." *International Zoo Yearbook* 7:120–22 (1967).

Sperry, Charles C. "Food Habits of the Coyote." *Wildlife Research Bulletin* 4. Washington, D.C.: U.S. Dept. of Interior, Fish and Wildlife Service, 1941. 70 pp.

Stenlund, Milton H. "A Field Study of the Timber Wolf (*Canis lupus*) on the Superior National Forest, Minnesota." *Minnesota Department of Conservation Technical Bulletin* 4, 1955. 55 pp.

Storm, G. L. "Movements and Activities of Foxes as Determined by Radio-Tracking." *Journal of Wildlife Management* 29:1–13 (1965).

Tanaka, Kojo. *Wildlife in Japan.* Tokyo: Yama-to-Keikoku Sha Co., Ltd., 1967.

Tate, George Henry Hamilton. *Mammals of Eastern Asia.* New York: The Macmillan Co., 1947.

Theberge, John B., and J. Bruce Falls. "Howling as a Means of Communication in Timber Wolves." *American Zoologist* 7:331–38 (1967).

Thenius, E. "Zur Abstammung der Rotwolfe (Gattung *Cuon* Hodgson)." *Osterreichische Zoologischer Zeitschrift* 5:377–87 (1954).

Tinbergen, Niko. "The Serengeti Research Project." *Animals* 9:28–35 (1966).

*Van Wormer, Joe. *The World of the Coyote.* Living World Books. Philadelphia: J. B. Lippincott Co., 1964.

Vesey-Fitzgerald, Brian. *Town Fox, Country Fox.* London: André Deutsch, 1965.

Vincent, Robert E. "Observations of Red Fox Behavior." *Ecology* 39:755–57 (1958).

Walker, Ernest P. *Mammals of the World,* Vol. II. 2d ed. Revised by John L. Paradiso. Baltimore, Md.: Johns Hopkins Press, 1968.

Walsh, John. *Time Is Short and the Water Rises.* New York: E. P. Dutton, 1967.

Wayman, Stan. "A Family of White Wolves." *Life,* October 16, 1967, pp. 34–44.

Wipf, Louise, and Richard M. Shackelford. "Chromosomes of a Fox Hybrid (*Alopex-Vulpes*)." *Proceedings of the National Academy of Sciences,* 35:468–72 (1949).

Wood, J. E. "Age Structure and Productivity of a Gray Fox Population." *Journal of Mammalogy* 39:74–86 (1958).

Woolpy, Jerome H., and Benson E. Ginsburg. "Wolf Socialization: A Study of Temperament in a Wild Social Species." *American Zoologist* 7:357–63 (1967).

Wyman, Jon. "The Jackals of the Serengeti." *Animals* 10:79–83 (1967).

Young, Stanley P., and Edward A. Goldman. *The Wolves of North America.* 2 vols. New York: Dover Publications, 1964. (First published in 1944 by the American Wildlife Institute, Washington, D.C.)

———., and H. H. T. Jackson. *The Clever Coyote.* Harrisburg, Pa.: The Stackpole Co.; and Washington, D.C.: Wildlife Management Institute, 1951.

Index

Abyssinian red fox. *See* Semyen fox
Africa. *See* specific animals
African hunting dog, 224-36, 260, 261
African sand fox, 126, 168-69
African wild dog. *See* African hunting dog
Aguará guazú. *See* Maned wolf
Aguarachaí *See* Field fox
Alaska: gray wolf in, 26, 27, 29, 43, 52, 57; red fox in, 136, 137, 141, 147-48
Algonquin Park, gray wolf in, 42, 43
Alopex lagopus. *See* Arctic fox
Anatomical charts, 251-55
Andean wolf. *See* Culpeo
Antarctic wolf, 192
Antelope (*See also* Thomson's gazelle): African hunting dog and, 229-30; jackals and, 91-94
Anubis (god), 88
Arctic fox, 173-81; chromosome number, 180, 262; red fox crosses, 264
Argentine fox. *See* Chilla
Arthritis, 30
Asian jackal. *See* Golden jackal
jackal
Ass fox. *See* Cape fox
Atelocynus microtis. *See* Small-eared dog
Atok. *See* Peruvian fox
Australia, 21 (*See also* Dingo); red fox in, 127-28, 139
Australian sheep-dog, 107
Azara's fox. *See* Forest fox
Baboons, jackals and, 97
Badgers, coyote and, 78
Baluchistan fox. *See* Hoary fox
Basenji, 115, 117
Bat-eared fox, 247-50, 260
Bates, Marston, 245-46
Beagles, 119
Bears: Arctic fox and polar, 178, 181; and red fox, 141; and wolves, 56-57
Beaver, gray wolf and, 32
Bengal fox, 126, 161-62
Bibliography, 265-70
Big-eared fox. *See* Bat-eared fox
Birds: Arctic fox and, 177-78; jackals and, 91; of prey (*See* specific prey); red fox and, 137-38
Bison, gray wolf and, 40-41
Black-backed jackal, 21, 89-97; crosses, 263-64; taxonomy, 257
Black-eared fox. *See* Bat-eared fox
Blanford's fox. *See* Hoary fox
Bobcats, red fox and, 141
Bounty system, 65-66
Brazil. *See* South American wild dogs
Bronze Age, 109, 110, 258
Brush wolf. *See* Coyote
Bukharian fox. *See* Hoary fox
Bulldog, 116-17
Burrows, Roger, 145, 147, 151
Bush dog, 243-46, 260, 261; taxonomy, 259

Cabrera, Angel, 191, 192, 199, 214
Canaan Dog, 114
Canada: Arctic fox in, 173, 176; gray wolf in, 26, 27
Cancer, 30
Caninae, 260, 261
Canis, 19-119; *adustus* (*See* Side-striped jackal); *aureus* (*See* Golden jackal); *dingo* (*See* Dingo); *familiaris* (*See* Domestic dog); *hallstromi*, 257; *latrans* (*See* Coyote); *lupaster*, 257; *lupis* (*See* Gray wolf); *mesomelas* (*See* Black-backed jackal); *niger,* 257; (*See* Red wolf); *rufus* (*See* Red wolf)
Cape fox, 126, 169-72
Cape hunting dog. *See* African hunting dog
Caribou, gray wolf and, 27, 32, 33, 38
Carrion. *See* specific animals
Cattle. *See* specific predators
Cerdocyon, 259; *thous* (*See* Forest fox)
Ceylon, golden jackal in, 85-86, 89
Chacolillo. *See* Gray fox
Cheetahs, jackals and, 93-94
Chilla, 199-201
Chiloé fox, 201
Chow-type dogs, 115
Chromosome numbers, 262. *See also* specific dogs
Chrysocyon brachyurus. *See* Maned wolf
Color, 16. *See also* specific dogs
Common fox. *See* Pampas fox
Common jackal. *See* Golden jackal
Compound 1080, 83

Corsac fox, 125, 157-61
Coy-dogs, 83, 263
Coyote, 20, 72-83; crosses, 83, 263; and kit fox, 157; and red fox, 141; and red wolf, 70-71
Crab-eating fox. *See* Forest fox
Crisler, Lois, 47, 176-77
Crosses, 262-64. *See also* specific dogs
Culpeo, 193-95
Cuon alpinus. See Dhole
Curled tails, 115
Custer wolf, 65
Cyanide gun, 82, 157
Cyon alpinus. See Dhole

Darwin, Charles, 45, 201
Dates, fennec and, 185
Deer: coyote and, 76-77; gray wolf and, 39-40, 41,42
Delalande's fox. *See* Bat-eared fox
Denmark, 108-9
Desert fox. *See* Fennec
Dhole, 237-42, 260, 261; taxonomy, 259
Dingo, 21, 99-107; and domestic dog, 106-7, 113, 117; taxonomy, 257
Disease. *See* specific dogs, illnesses
Distemper, 29, 234-35
Domestic dogs, 19, 21-22, 107-19; anatomical charts, 251-55; chromosome number, 262; and coyote, 83, 263; crosses, 262-64 (*See also* specific dogs); dingo and, 106-7, 113, 117; and gray wolf, 21, 57-59, 111, 112-13, 116-18, 263; taxonomy, 257-58
Dung beetles, 249
Dusicyon, 191-203, 259; *australis*, 192; *culpaeolus* 195; *culpaeus*, 193-95; *fulvipes*, 201; *griseus*, 199-201; *gymnocercus,* 195-99; *inca,* 199; *sechurae; vetulus,* 202-3

Eagles, 123, 141-42
Ears, 16 (*See also* specific dogs); lop, 115
Eaton, Randall, 93
Egypt: domestic dog in, 109; jackals in, 88
Egyptian jackal, 257
Encephalitis, 181, 190
England (the British): and gray wolf, 27; and red fox, 126ff., 136, 139, 141, 142
Ethiopia, 120-23
Ethiopian red fox. *See* Semyen fox
Europe. *See* specific dogs
European wolf. *See* Gray wolf

Falkland fox; Falkland island dog; Falkland wolf, 192
Feet, 16. *See also* specific dogs
Fennec 182-85; chromosome number, 262
Field fox, 202-3
Finland, gray wolf in, 27, 57
Fish: gray wolf and, 32; raccoon dog and, 220-21
Fleas, 81-82, 96, 143
Forest fox, 204-7; domestic dog crosses, 207, 264
Foxes, 57, 120-211. *See also* specific kinds
Fuller, John L., 112-13, 117, 119
Fur, 16, 18. *See also* specific dogs

Genetic information, 262-64
German shepherds, 58
Giant fox. *See* Maned wolf
Gilfillian, Archer, 74
Goats. *See* specific predators
Golden jackal 21, 83-89; chromosome number, 262; crosses, 263; and domestic dog, 111-12, 263; taxonomy, 257
Grass, foxes and, 138
Gray fox, 186-90; South American (*See Dusicyon*)
Gray jackal. *See* Golden jackal; Side-striped jackal
Gray Rhodesian jackal. *See* Side-striped jackal
Gray wolf, 19, 20, 22-67; and coyote, 80-81; and domestic dog, 21, 57-59, 111, 112-13, 116-18, 263; and red fox, 138, 141
Greyhound-type dogs, 109, 111, 115
Guará. *See* Maned wolf
Guarachaim. *See* Field fox

Hair, 16. *See also* specific dogs
Hedgehogs, red fox and, 137
Heeler dog, 107
Hoary fox, 126, 163; South American (*See* Field fox)
Hookworm, 81
Horses, wolves and, 60
Hough, 1948 study by, 261
Hunting, 17, 18. *See also* specific predators, prey
Hyena dog. *See* African hunting dog
Hyenas, African hunting dog and, 234

India: Bengal fox in, 161-62; golden jackal in, 85, 89; gray wolf in, 61-62; red fox in, 136
Indian fox. *See* Bengal fox
Indian jackal. *See* Golden jackal
Ireland, gray wolf in, 27
Irish wolfhound, 63
Iron Age, 110
Isle Royale, 26, 33, 37-38, 41ff., 50, 56, 57, 80, 138, 141
Israel, pariah dogs in, 113-14
Italy, gray wolf in, 28

Jackals, 20-21, 83-99 (*See also* specific kinds); crosses, 262-63, 263-64 (*See also* specific dogs); taxonomy, 257
Jaguapitanga. *See* Field fox
Jaguars, coyote and, 81
Japan, raccoon dog in, 217
Jericho, 109

Kama fox. *See* Cape fox
Kangaroos, 103
Kelpie, 107
Kidney nematodes, 216
Kit fox, 125, 153-57; taxonomy, 258-59
Kühme, Wolfdietrich, 228, 232

Lake dog, 258
Lapland, gray wolf in, 27
Leakey, Louis, 249-50
Lemmings, 174-76, 177, 178

Leopards, jackals and, 97
Lions, jackals and, 89, 91
Litters, 17. *See also* specific dogs
Little gray fox. *See* Chilla
Livestock. *See* specific predators
Lobo. *See* Gray wolf
"Lobo" (renegade wolf), 65
Lobo crinado. *See* Maned wolf
Lone wolves, 50
Long-eared fox. *See* Bat-eared fox
Lop ears, 115
Lorenz, Konrad, 112
Lycaon pictus. See African hunting dog
Lynxes, 57, 141

Maglemosian culture, 109
Maned wolf, 212-16; taxonomy, 259
Mange, 82, 143
Matthew, 1930 study by, 260
Mech, L. David, 37
Mesopotamia, 109
Mexico, gray wolf in, 26
Mice: corsac fox and, 159; jackals and, 91; red fox and, 136-37
"Mobbing," 138
Moles, 137
Moose, gray wolf and, 32ff., 41, 50
Mountain sheep, gray wolf and, 32, 39
Murie, Adolph, 53, 147-48
Musk-ox, gray wolf and, 32
Muzzles, 16. *See also* specific dogs
Myxomatosis, 100, 136

Names, scientific, 256-57
New Guinea singing dog, 257
North America. *See* specific dogs
Northern wolf. *See* Gray wolf
Norway, gray wolf in, 27
Novikov, G. A., 163
Nyctereutes procyonoides. See Raccoon dog

Otocyon megalotis. See Bat-eared fox
Otocyoninae, 260

Pakistan, red fox in, 136
Pampas fox, 195-99; domestic dog crosses, 264
Paraguayan fox. *See* Pampas fox
Pariah dogs, 21, 113-14, 117
Peat dog, 258
Peruvian fox, 199
Piercer, the, 62-63
Pocock, R. I., 163, 164, 241
Poison. *See* specific animals, poisons
Polar bear, Arctic fox and, 178, 181
Posture, 16. *See also* specific dogs
Prairie wolf. *See* Coyote
Pumas, coyote and, 81
Pups, 17. *See also* specific dogs
Python, jackals and, 97

Rabbits and hares: coyote and, 76, 77, 81; dingo and, 100; red fox and, 136
Rabies, 29-30, 81, 123, 143, 181
Raccoon dog, 217-23; chromosome number 262
Rats, Semyen fox and, 121-23
Ravens, wolves and, 57
Red dog. *See* Semyen fox
Red fox, 125, 126-53; Abyssinian (*See* Semyen fox); Arctic fox crosses, 264; chromosome number, 262; Ethiopian (*See* Semyen fox); South American (*See* Culpeo); taxonomy, 258
Red jackal. *See* Semyen fox
Red River dog, 107
Red wolf, 20, 67-72 (*See also* Maned wolf; Semyen fox); and coyote, 70-71, 263; taxonomy, 257
Regurgitated meat, 54-55. *See also* specific dogs
Reindeer, gray wolf and, 60
Renegade wolves, 64-65
Rodents, 18. *See also* specific animals
Roman times, 111
Round worm, 81
Rue, Leonard Lee, 147
Rüppell's sand fox, 126, 165-68; chromosome number, 166-68, 262
Russian borzoi, 63

Saluki, 109
Samoyed, 58
Santa Elena Fox, 195
Savanna fox. *See* Forest fox
Scent glands, 124-25. *See also* specific dogs
Scientific names, 256-57
Scott, J. P., 112-13, 117,119
Sechura fox, 203, 259
Semyen fox 120-23; taxonomy, 258
Serengeti Plains (Tanzania), 84, 86-88, 91-93, 94
Seton, Ernest Thompson, 65
Sheep. *See* specific predators
Shrews, 137
Side-striped jackal, 21, 97-99; taxonomy, 257
Silver-backed jackal. *See* Golden jackal
Silver jackal. *See* Cape fox
Simenia simensis. See Semyen fox
Simocyoninae, 260-61
Small-eared dog, 208-11; taxonomy, 259
Small-toothed dog. *See* Field fox
Sodium fluoroacetate, 83
Sounds. *See* specific dogs
South American wild dogs, 191-216, 243-46; taxonomy, 259
Soviet Union. *See* specific dogs
Speothos panamensis. See Bush dog, 259
Speothos venaticus. See Bush dog
Sperry, Charles, 76, 77
Squirrels, 136, 137
Steppe fox. *See* Corsac fox
Stone Age, 109, 110
Strychnine, 82, 106, 153
Sweden, gray wolf in, 27
Swift fox, 153, 258-59

Tails, 16 (*See also* specific dogs); curled, 115
Tanzania, 84, 86-88, 91-93, 94
Tapeworm, 81, 142
Tartar fox. *See* Corsac fox
Tasmanian wolf, 100
Taxonomy, 257-59
Teeth, 16, 260-61. *See also* specific dogs

Terriers, 118-19, 264
Thomson's gazelle, 86-88, 93
"Three Toes" (renegade wolf), 58-59
Tibetan sand fox, 126, 163-65
Ticks, 96, 143
Tigers:, golden jackal and, 89; wolves and, 56
Timber wolf. *See* Gray wolf
Tinbergen, Niko, 227-28
Trapping, 63-64. *See also* specific animals

United States. *See* specific dogs
Urocyon cinereoargenteus, 186-90

Vocies. *See* specific dogs
Voles, 136-37, 159, 177
Vulpes, 124-72; *bengalensis* (*See* Bengal fox); *cana* (*See* Hoary fox); *chama* (*See* Cape fox); *corsac* (*See* Corsac fox); *ferrilata* (*See* Tibetan sand fox); *fulva* (*See* Red fox); *macrotis,* 258, 259; *pallida* (*See* African sand fox); *rüppelli* (*See* Ruppell's sand

Warrigal. *See* Dingo
Wildebeests, African hunting dog and, 230-31
Wolverines, 57, 141
Wolves, 19-20, 22-72. *See also* specific kinds

Yellow jackal. *See* Golden jackal
Yellowstone Park, 26
Yepes, José, 191

Zebra, black-backed jackals feed on, 92